I0488061

THE THEORY OF PRIMES

THE THEORY OF PRIMES

▼

Mark Meek

Writers Club Press

San Jose New York Lincoln Shanghai

The Theory of Primes

All Rights Reserved © 2000 by Mark Meek

No part of this book may be reproduced or transmitted in any form or by any means, graphic, electronic, or mechanical, including photocopying, recording, taping, or by any information storage retrieval system, without the permission in writing from the publisher.

Writers Club Press
an imprint of iUniverse.com, Inc.

For information address:
iUniverse.com, Inc.
5220 S 16th, Ste. 200
Lincoln, NE 68512
www.iuniverse.com

ISBN: 0-595-16417-X

Printed in the United States of America

This book is dedicated to the God of the Bible and to his only begotten Son, Jesus Christ. No matter what human beings may discover, we are only finding what God created long ago. Let's give credit where credit is due.

INTRODUCTION

▼

There is a level of the physical universe more fundamental and primal than anything yet explored. Following an exhaustive search, it seemed it was yet to be pointed out that there is a simple formula, a sequence, by which everything that can possibly be exists and operates. I have named this formula or sequence "The Primes", because it is the most primary component of physical reality.

Before the Big Bang brought the universe into existence, something had to exist to define what could possibly exist and could happen. This is the purpose of the primes and I believe that God set down the primes before initiating the Big Bang to get the universe under way. If a constitution is what we call a written plan of government, we could say that the primes are the constitution of reality; encompassing all that exists, all that has ever existed, all that ever will exist or could possibly exist. Everything from the Big Bang onward has fallen into this defining sequence.

The Theory of Primes is an entirely new field. There is nothing to quote or refer to and no reason to do so. I may be just a little bit biased but I think the exploration and mapping of the primes is one of the fundamental discoveries ever made by a human being. I have long been passionate about science, particularly astronomy and physics, and it would fulfill a lifelong dream to really contribute something myself.

I came across the primes somewhat accidentally. I had noticed that no one seems to have ever categorized the various patterns which are manifested in the universe and the world around us as entities in themselves, rather than as a part of whatever topic being dealt with. I spent several months puzzling over a way to list and categorize these patterns into a logical inter-related sequence.

One fine day a flash of insight revealed to me that any possible pattern could invariably be broken down into a four-part sequence. Everything in existence conformed to this sequence. It looked as if I had found the primes. I am in the habit of asking God for wisdom and I do not doubt that this discovery is the result.

The primes must be manifested to have any real existence in the same way as all mathematical entities such as numbers or spheres. However, it can be shown that primes are more even more fundamental than the everyday counting numbers. This means that the assumption of simple numbers being the most fundamental unit of mathematics is untrue.

Down through the centuries, as mathematics got more and more complex, there was always the hidden branch that was the most fundamental of all. This branch is the primes and has not been explained until now. Mathematics, since the inclusion of primes, now has the capability not just to quantify and describe what exists but to define what can exist.

CHAPTER ONE

▼

PRIMES

THE SEARCH FOR PRIMACY

How far can we get when we try to break down the world around us into it's core components, what we could call primacy or finding that which is most primal? The ancient Greeks thought about it and decided that everything could be reduced to four basic elements; fire, air, water and, earth. Everything of a mechanical nature can be broken down into a few simple machines such as levers and pulleys. Likewise, there are only a few primary geometric shapes. Prime numbers are defined as those which can be divided by no numbers other than itself and one.

Chemists talk about the elements of the periodic table as if these different kinds of atoms are the most basic and primal of all things. Indeed the Greek word "atomos" means "that which cannot be cut". Physicists usually take things a step deeper to the realm of quarks or space-time or the four basic forces operating the universe. These forces are gravity, the electromagnetic force, the strong nuclear force binding atomic nuclei together and, the weak nuclear force associated with radioactivity. Many people believe that gravity is not actually a force but rather an innate property of space itself.

When we analyze physical reality down into its core components, we talk of space, time, matter, elements, energy and, forces. I believe that there is an even deeper, more fundamental level. The purpose of the Theory of Primes is to go even deeper, to this most primal level of physical reality. The Theory of Primes could also be called "Primary Theory".

I may possibly be a little bit biased but I believe that this is one of the few fundamental discoveries ever made. I have long been interested in science, particularly physics and astronomy. I recall once thinking about what an incredibly orderly place the universe was, from the most distant galaxies to sub-atomic particles, from the electromagnetic spectrum to the basic forces that govern the behavior of everything.

In fact, the universe seemed to me to be even more orderly than it should be, considering the known facts of science and mathematics. Maybe there was something else out there contributing to the orderliness and organization of the universe that no one had yet uncovered.

I sensed that I might possibly be onto something but I could not nail it down. I though hard about what it could be that was contributing to this excessive orderliness that seemed to pervade everything. But whatever it was, it was as elusive to me as the shimmering water mirage that can often be seen on the road up ahead on a hot summer day, until you reach where it seems to be only to find that it has moved further back.

I pretty much let the though go that there was something else out there, as yet undefined, organizing or operating the physical reality of the universe. Particularly something that I could get hold of just by wondering about it.

One day many years later, an incredible thought flashed into my mind. It was not quite like finding God but it was very significant. I realized that there are only four basic things that can exist or can happen. Everything that we have ever known or ever will know consists of manifestations of and combinations of these four things. I decided to name these four entities primes because they exist at the most primary or primal level of physical reality. The Theory of Primes that I have developed explains why the universe in all it's realms is more orderly than it often appears that it should be, considering the random free-for-all chaos of the laws of nature alone. Throughout this book, we will introduce postulates periodically.

POSTULATE 1: PRIMES DEFINE WHAT CAN EXIST AND WHAT CAN HAPPEN. EVERYTHING IS MANIFESTATIONS OF AND COMBINATIONS OF THE PRIMES.

WHAT ARE PRIMES?

Primes could be described as the forms or patterns that are manifested by everything that does exist or can exist. Primes are on a much deeper level of physical reality than atoms or their components. Up to now, it seemed as if the dimensions are the most primal component of the physical universe, the enclosure of space and time within which everything exists. Just as there are four primes governing physical reality, there are four dimensions doing the same, three dimensions of space and one of time.

But the primes are at a deeper level than dimensions and, are no relation to the dimensions. The dimensions did not exist until the big bang brought space and time and apparently everything else into existence. The primes existed before the big bang and governed the physical progress of the big bang.

When we understand primes, we think of the universe not as matter, forces and, energy operating in the framework of the four dimensions. But as all of these things, including the dimensions, manifesting the four primes. It is possible to imagine a universe of more than four dimensions, science fiction writers have done it many times. However, there are only four primes and I cannot imagine what a fifth one could be. No matter how many dimensions there were, or whether there was an infinity of dimensions, the four primes would remain the same.

To understand what primes are, we must be careful to separate primes from that which is a property of space-time, matter, forces and, energy. All of which I will henceforth refer to as "the given". We are "given" the material, physical universe but the primes had to exist prior to the big bang, which is believed to be the birth of the given. Although the primes were not manifested until the advent of the given.

As a general rule, if it did not exist before the big bang it has nothing to do with primes, it is only a manifestation of the primes. No changes in the given has any effect whatsoever on the nature of the primes. We can deal with atoms or radiation or economics or sports and the primes remain the same. This is why primes are called primes, for being more primal, more primary, than whatever we are dealing with, whatever it is that is manifesting the primes.

POSTULATE 2: THE GIVEN MANIFESTS PRIMES

Just as numbers existed before the big bang, before there was anything to count, the primes also were there, waiting to be manifested by the "stuff" introduced in the big bang, the beginning of the universe. If this "stuff" had been entirely different, the primes as well as the numbers would be manifested just the same. No matter what exists, as long as there is any form of reality, the same four primes will be there as manifestations in that reality. There is an infinity of numbers but, there is only four primes.

We do not need scientific instruments such as telescopes, microscopes or, space probes to discover the primes any more than we need such instruments to discover the dimensions. Einstein developed his theory of relativity, and Newton his theory of gravity, merely by thought. Whether we look at things on the scale of microns or kilometers or light years, we see the same primes as well as the same dimensions providing an orderly framework for things to exist and events to happen.

THE GAP BETWEEN SPIRITUAL AND PHYSICAL REALITY

I should state right now that I believe very much in the God of the Bible as the initiator of the big bang and, therefore the creator of the universe. However I did not believe that God reaches down and guides the motions of every atom all the time. God set the universe in motion and on the whole left the physical universe at that.

This is not a religious book, it is a fundamental scientific theory. I do think that primes were the first creation of God, providing the framework for the big bang and everything that followed. Before God decided what would exist, he must have thought of some kind of rules or patterns for what could be permitted to exist in the physical universe. These rules or patterns are called primes.

On one side we have what we know of the physical universe; space, matter, energy and so on. On the other side, we have the religious and philosophical meaning of the universe. Both sides have been discussed and written about for thousands of years.

What if there is a gap between these two sides? What might we find in this gap? What I believe would be found there are the primes. The four primes are entities that are not in the spiritual realm but are at the deepest level of the non-spiritual, non-philosophical realm.

POSTULATE 3: PRIMES ARE THE MOST FUNDAMENTAL COMPONENT OF PHYSICAL REALITY

We could compare the primes to our own lives. There are religious and philosophical reasons for why we exist on one side. On the other side, biologists and doctors can tell us all about the physical workings of our bodies. In between these two sides we have our daily lives, we go to work or school, read, exercise, think, reproduce, buy things, invent and discover things and, so on.

Most of the details of daily life are neither spiritual or philosophical nor are purely physical. On an ordinary level, we exist simply to do things. The primes can likewise be considered as the level of reality involved with existing and "doing things" without going as far as considering the material components that actually "do" the things, the physics, chemistry or biology involved.

THE WALLS AND CORRIDORS OF REALITY

My theory is that the primes serve to set up walls and corridors to govern the operation of reality. No matter what that reality is.

An analogy to help understand primes as the walls and corridors of reality is a city. Suppose you were going to drive around a city or town. It seems as if you can go wherever you want but actually, you must follow the existing street and parking lot pattern don't you? You must follow the traffic signs and signals. You cannot just drive in a straight line wherever you feel like it.

What if you are at home? You can go wherever you want in your own house can't you? Actually, you must follow the predefined pattern of rooms in the house. If you just walk in a straight line, you will probably run into a wall. You can go wherever you want but it must be within the patterns set up in the form of walls in a house or streets in a city.

So it is with primes. Anything is possible that the given space, time, matter and, energy in the universe make possible and that the laws of physics will allow but it must be done within the patterns, the walls and corridors, set up in the form of primes. This is why the universe seemed to me to be more orderly and organized than it should be, considering the known facts of science without the primes.

If the given is compared to the cast thrust onto a stage to put on the grand spectacle we call the universe, the primes would be the stage.

If the given is compared to a concrete structure, the primes would be the forms used when the concrete was set.

If the primes are compared to a deck of cards, the given is the particular card game being played.

If the primes are compared to a jar, the given is whatever the jar is filled with.

MATHEMATICS AND MANIFESTATIONS

Primes can be considered as a branch of mathematics more primary than even the integers, the common counting numbers. We have seen that primes and mathematical entities such as numbers both existed before the big bang but could not be manifested until some kind of "stuff" existed such as a universe or a six-pack of Coca-Cola. A number, like a prime, is something that must be manifested in order to become real.

Consider the number 6 for example. There is no such thing as the number 6 which exists. We cannot see or photograph the number 6, we cannot visit it like we can the Eiffel Tower, it is not an entity in itself. The number 6, or any other number, must be manifested to become real. When we are in the store and see a six-pack of Coca-Cola, then the number 6 becomes real because it has been manifested.

A number by itself is meaningless until it is manifested. Let's take the number 17.32 as an example. What does this number mean? 17.32 miles? 17.32 grams? 17.32 light-years? It is only when the number is manifested by some given (miles, liters of olive oil, tons of wheat, etc.) that it means anything. The same applies to all branches of mathematics, it must be manifested to be real. A square, a triangle, a sphere or, a cube is not real until it is manifested.

Primes, like numbers, do not become real until manifested. Primes, however are more primal than any existing arithmetic or mathematics. All branches of mathematics describe and quantify what exists or what happens, primes set the patterns for what can possibly exist or happen.

POSTULATE 4: PRIMES ARE MATHEMATICAL ENTITIES WHICH ARE MADE REAL BY MANIFESTATION.

Present mathematics follows what is already existent, it quantifies the given of the physical universe; the dimensions, matter, energy, people etc. Primes lead the way by setting the patterns of what can exist or happen with the given of the physical universe. Primes define and set the patterns of what can exist or can happen. The relationship happens in the following sequence.

1) Primes define and establish the patterns of what can exist and what can happen.
2) The given comes into existence; things exist and events happen.
3) Conventional mathematics is developed to describe and quantify what exists and what happens.

Two actions on the four basic primes compose everything that can exist. These two actions are manifestation and combination. Primes must be manifested to become real and what actually exists depends on the given that manifests the primes. In our case; space, time, forces, matter and, energy is the given. The four basic primes can be combined in an infinity of possible ways that is limited only by the nature of the given that is manifesting the primes.

The reality that we know is the property of both the primes and the given. Just as the reality of a jar of marmalade is the property of both the jar and the marmalade. With primes, anything is possible and primes never run out. The possibilities and limitations of what actually does and can happen depends on the given and it's qualities. This explains why something so fundamental to the nature of reality as the primes has not been explained until now.

Next, lets meet the four primes one by one.

CHAPTER TWO

▼

MEET THE PRIMES

THE FIRST PRIME: DOMAIN

The first and most primal of the primes is the domain. It is impossible to have anything without having a domain. Anything that has a boundary of any description is a domain. Anything that acts upon something or is acted upon is a domain. Anything that can be measured or has a center or a focal point is a domain. Anything that can be defined or located involves a domain.

A bank account, an atom, the universe, the Baroque period, a screwdriver, an ocean, a window of opportunity, February 8, 1965, that which is possible, that which is impossible, a planet, a sheepdog, a magnetic field, a watershed, Marcia on the Brady Bunch, a library, a nation, holiness, evil, the nineteenth century, a soccer play, a cloud, dominance, relevance, irrelevance, poverty, a city and, the Arabic language are all domains.

All are equal manifestations of the first prime. The easiest way to define a domain is simply as a condition that exists. Usually in our spatially-oriented universe, it is a condition that exists at one point but not at some other point. All opposites, such as possible and impossible are domains.

Domains can be defined in terms of what makes it a domain or what it is made of. The earth is a domain consisting of various rocks and minerals and made into a domain by gravity. Domains are usually defined by some kind of boundary, one side of the boundary is within the domain, the other side is not. In our given, our universe, the simplest boundary is often space itself, distance. Other boundaries that we would be familiar with defining a domain would be fronts, fault lines, gravitational fields and, magnetic fields. A boundary defining a domain may be sharp and well-defined or, gradual and ill-defined. But, the domain would be just as much a manifestation of the first prime, the domain.

Any event in our universe, anything that happens, produces time domains in the form of before and after. Since time is a part of our given, domains can be in time as well as in space. When something has a beginning or an end, we are considering a domain in time. When we say that "she was walking and listening to the radio", we are speaking of a time domain containing both walking and listening to a radio. A millenium is a familiar example of a time domain.

Generally, anything that is a part of our given produces one or more domains. Anytime that something has anything in common with something else, a domain of some kind is produced. Anything which we can give a name to or, can put "a", "an" or, "the" in front of is a domain.

POSTULATE 5: THE FIRST PRIME IS DOMAIN AND IS MANIFESTED WHENEVER ANYTHING EXISTS.

Domains can consist of sub-domains and can be a part of larger domains. In our given, our universe, this is almost always the case. Matter as we know it can be broken down into particles such as quarks and electrons. In magnetism, a domain is a particle of the metal which acts as a tiny magnet.

Consider a house made of bricks. We can look at the entire house, we can move closer and look at one brick or, we can zoom in still closer until we are looking at one ceramic grain in the brick. We could keep zooming in until eventually we were looking at one atom in one grain in one brick in the house. What we have here is four completely different domains existing concurrently on different levels. Each of the domains are sub-domains and is just as legitimate an example of the domain prime, just as much a manifestation of domainess, as the others.

In dealing with domains made of sub-domains, we can define that domain on the level that we are dealing with as the dominant domain. It does not mean that the other domains existing concurrently are any less a manifestation of domainess, the definition is only for our convenience. If you roll a ball bearing (kinetic energy), the dominant domain is the ball bearing. If you heat the ball bearing, the dominant domains are the atoms in the ball bearing since heat is kinetic energy of atoms and molecules.

THE SECOND PRIME: LEVEL

The existence of the first prime, domain, inevitably brings into existence the second prime, which is level. We have already seen how a domain can exist on different levels as the larger domain or it's sub-domains. We should make the name a plural, levels, rather than singular because if only one level could exist, the idea of level as a prime would be meaningless. If there was only a single infinitesimal speck of a domain in existence, we could do without the concept of levels. But anything more than that brings levels into manifestation.

Levels simply involve a difference or an amount of some kind. Whenever there are differences of any form, levels are manifested. Domains tend to be a simple yes or no issue, either a domain exists or it does not. Levels tend to be in infinite number depending on how finely we want to measure. We could say that domain is digital because it either exists or it does not while level is analog because it exists as a range which can be divided infinitely.

The difference between two levels is a level in itself, we could say that of all the planets in the solar system, pluto has the greatest contrast between the light and dark areas on it's surface. This means that pluto has the highest level of difference between it's surface domains of a high level of light and it's surface domains with a low level of light.

The earth is a domain. As we move further from earth, its gravitational field, as well as its apparent angular diameter in the sky decreases. Two levels, the distance from earth and the potential energy of the body resulting from its altitude increase. Two other levels, the earth's gravitational field and, it's apparent angular diameter decrease at the same time.

Is the earth flat or spherical? Your backyard is probably more or less flat. Yet, your backyard is on the earth and we can definitely see in the photographs taken from space that the earth is spherical. How can this be? Because we are considering the earth on different levels.

Levels can involve an amount rather than a difference. An amount is something at a certain level which could potentially be different but is not. A difference implies a comparison while an amount does not.

All questions of direction and location are issues of levels, we can describe any location by expressing it's level of northness or southness as well as eastness or westness and upness or downness. Steering is a levels issue. Navigation deals with levels such as latitude and longitude. Speed and velocity are both levels, the difference being that speed is a "one dimensional level" while velocity, which gives direction as well as speed, is a multi-dimensional level. A graph plots a domain out of levels of points.

The concept of a point in mathematics is an infinitesimal one-dimensional domain of zero real dimension, which is not a domain at all in the real world but, is an issue of level in that it's significance lies solely in it's location. The only way to have a domain manifested without levels is to have such an infinitely tiny speck as the only thing in existence. But, a point exists only as a mathematical entity. In the real universe, domains manifest levels.

We call levels the second prime because the first prime, domain is necessary to manifest levels. Aside from our mathematical point, there can be no domains without levels and we know that there can be no levels without domains. A bank account is a domain, the amount of money in the account is a level. A city is a domain, directions or location is levels. A geometric figure is a domain, it's dimensions are levels. The Arabic language is a domain, the number of people that can speak it is a level. salt crystals dissolve in water as sub-domains, the resulting salinity of the water is a level. The universe is a domain, the distances between various celestial bodies are levels. An ocean is a domain, the amount of water actually in it is a level. A skyscraper is a domain, each floor is a level. Energy is a level, a type of energy is a domain. Numbers are levels, odd numbers are a domain.

The different domains that we have described in the brick house are all domains, the difference is in the levels. Whenever we have any form of amounts or differences, we have levels. Adjectives or descriptions are just as much manifestations of levels as are numbers. Adjective levels are often expressed in terms of domain rather than level. Nationality, Taiwanese for example, could be expressed as a level of association with Taiwan but, would probably be considered as a domain issue, a member of the domain known as Taiwan.

As with domain, the exact nature of a level depends on the given. The number of protons in an atom is a level. The beer-drinking capacity of Joe Smith is just as much a manifestation of levelness. Whatever the given, it is practically impossible to have domains without levels. Level is a prime but, as with domain, the relevant categories of levels are dependent on the given. In our universe, objects attract each other due to gravity while like electrical charges repel and unlike electrical charges attract and the force thus generated is a manifestation of levels. However if there was a universe where this was not the case, level would be manifested by whatever rules did exist in that given.

Ironically, one issue that helps us to understand primes is that primes themselves do not manifest levels. This means that no domain is more or less a manifestation of the domain prime than any other domain and no set of levels is more or less a manifestation of the level prime than any other set of levels. Beauty is a description that can be expressed in various levels and that must be manifested like primes or numbers, we do not see beauty just floating around by itself unless it is manifested. However, beauty is not a prime, one way in which we know this is that we can attach levels to it.

POSTULATE 6: THE SECOND PRIME IS LEVEL AND IS ALWAYS MANIFESTED BY DOMAIN. AN EXPRESSION CONCERNING DOMAIN IS LEVEL.

A prime, such as domain, cannot have another prime, such as levels, attached to it. Only the manifestation of domain can be influenced by level or any of the other primes. One domain may be larger or hotter than another domain and that is what brings levels into existence in our given. But no domain has any more domainess than another. It is only the manifestation of domain which we can attach levels to, not the domain prime itself.

We may consider one domain as more obvious and sharply defined than another but that is due to the nature of the given which is manifesting the domain prime, not the prime itself. This can also be seen in numbers, a dozen of something is just as much a manifestation of a dozen as a dozen of anything else. This shows the similarities between primes and conventional mathematics. Whenever you have six of anything, no set of six is any more a manifestation of six than any other. No sphere of equal spherical perfection is any more a manifestation of sphereness than any other. In the same way, no domain is any more a manifestation of domainess than any other domain and no level is any more a manifestation of levelness than any other level.

POSTULATE 7: PRIMES CANNOT EXIST IN LEVELS. NO MANIFESTATION OF ANY PRIME IS ANY MORE OR LESS A MANIFESTATION OF THAT PRIME THAN ANY OTHER MANIFESTATION OF THAT PRIME. ONLY MANIFESTATIONS OF A PRIME CAN MANIFEST LEVELS.

If domain is defined by a boundary, then level is defined by a barrier. A barrier means something does not happen because the barrier is there that would happen if the barrier was not there. If you have seventeen dollars, there is a barrier to your purchase of a book that cost eighteen dollars. This is the essence of level just as the boundary is the essence of domain.

Barrier is a stronger term than boundary. The description "stronger" is of course, a level. A boundary simply defines a limit while a barrier enforces it. Of course, in our familiar world, the concept of barrier and boundary often merge together. In our universe, the simplest barrier, as well as boundary, is often space itself. In the skyscraper that we have described, we consider floors as levels only because there is a barrier, gravity, preventing movement between floors which requires some kind of link to be utilized, such as a stairs or elevator.

POSTULATE 8: DOMAIN IS DEFINED BY BOUNDARY WHILE LEVEL IS DEFINED BY BARRIER. A BARRIER REQUIRES SOME MANIFESTATION OF LEVEL TO CROSS WHILE BOUNDARY DOES NOT.

Since levels result from domains, we should not be surprised that the two display a close relationship. Domains in our given can, of course, be described in terms of levels such as volume or, diameter. Distance and direction are issues of both level and domain. The mathematical concept of a point is not a meaningful domain in itself but describes a location in some kind of domain. The electromagnetic spectrum is usually divided into domains, such as ultraviolet and radio waves, which are defined by levels of frequency. Units, such as meters and liters, are on the line where domain and level meet. Any unit is a domain but the very definition of a unit is by level.

One way of demonstrating the veracity of the Theory of Primes is that primes "fit right in" to the universe. If primes are indeed the definition of all that can exist, then we should see some kind of resemblance between the primes and the given, our universe. And we do, in the way that the two basic constituents of the physical universe, matter and energy, "blend into" each other in that matter and energy are interconvertible. The interconvertability of domain and level, which are the two fundamental primes is almost a mirror image of that between matter and energy.

We know that primes are a branch of mathematics more primal than the counting numbers themselves simply because numbers are levels and something must exist to manifest and give meaning to the counting numbers. This something will invariably be some manifestation of the first prime, domain. While numbers are a quantification of the second prime, level.

Hierarchies, layers and, energy are all levels. Thinking mathematically, anything we can put a number on is obviously a level, while anything we can attach a geometric shape or condition to is a domain. Anything we can attach a level to is a domain. Every domain comes with an attached domain of levels. The planet Jupiter for example, is a domain. Jupiter cannot exist just as a manifestation of domain. It brings into manifestation all manner of levels from it's diameter and rotational period to it's revolution period around the sun. Geometric shapes are domains that inevitably manifest levels such as eccentricity and measurements.

A period of time is a domain while age is a level. Relevance and irrelevance are domains that can be expressed in levels as are rich and poor. An object is a domain while it's mass and weight are levels. The three states of matter; solid, liquid and, gas are levels of energy in a domain.

Our language does not have the precision of mathematics and how a statement is worded has a great effect on the manifestation of primes that it implies. Take the word "heat" for example. When you move closer to a wood stove to keep warm, heat is a domain.

When you are waiting for water to boil, heat is a level. When you are applying heat to something, heat is a compensation.

Domains and levels are often manifested in each other. A weatherman talking of a warm air mass and a cold air mass is referring to domains. Even though the domains are defined by the level of temperature. When we use a term with both a level and a domain, one or the other may be dominant. A ton is a domain, twenty is a level, twenty tons is a level. Warm is a level, water is a domain, warm water is a domain. This is however, due to the inexactness of our language rather than the imprecision of primes.

The planets are sub-domains of the solar system while Kepler's laws of planetary motion define levels. A measurement of time or distance can be both a domain and a level, depending on how the measurement is used. A period of time in history is a domain, the time it takes for a cake to bake is a level.

POSTULATE 9: PRIMES MUST BE MANIFESTED IN SEQUENCE. A CONDITION IS A DOMAIN. A CONDITION OF A CONDITION IS A LEVEL. ANYTHING THAT A NUMBER OR DESCRIPTION CAN BE ATTACHED TO IS A DOMAIN.

Giving the primes in order, domain is the first prime and results in the existence of level, the second prime. In other words, domain must be manifested before level. But when we deal with domains composed of sub-domains, we find that levels associated with the sub-domains affect the larger domain. In our given, this can be easily seen in melting and freezing and other state of matter issues resulting from the kinetic energy of the atoms, which are the sub-domains of the larger domain.

To describe what a domain is made of in terms of atoms, the level of the number of protons in the domain's atoms will tell us what element(s) it is made of. Domains consisting of sub-domains and sub-sub-domains are of course on different levels. Domain is the only prime which can manifest sub-primes, there can be sub-domains but no sub-levels because any sub-levels would be manifested from sub-domains. Many different worlds exist in the same place on different levels. Primes themselves can be manifested on an infinity of levels in our universe. The only limit to the manifestation of primes is of course, the given and it's nature. Which to us in our universe, means the properties of space-time, matter and, energy.

POSTULATE 10: THE ONLY PRIME THAT CAN MANIFEST SUB-PRIMES IS DOMAIN. THERE CAN BE NO SUB-LEVELS BECAUSE DOMAIN IS THE FIRST PRIME AND ANY MANIFESTATION OF SUB-LEVELS WOULD NECESSARILY BE DUE TO SUB-DOMAINS.

THE THIRD PRIME: COMPARISON

The domain is the first prime and the basis of the other primes. The existence of at least one domain brings the second prime, levels into manifestation. The existence of levels brings the third prime, comparison into manifestation. Any time we have domain we have levels and any time we have levels, we have comparison taking place.

Every domain acts as an "antenna" seeking instruction on what it is to do. The "antenna" does this seeking by "measuring" levels, which is why comparisons do not happen until levels exist. The levels are "compared", hence the name. All levels are open to comparison. It is the domain which does the comparison.

This third prime, comparison, could be referred to as the "governor" of the universe. The domains and levels form and bring in comparison to keep everything in order. If the level of one thing is higher, the domain behaves in a certain way. If the level of another thing is higher, the domain behaves in another way.

POSTULATE 11: THE THIRD PRIME IS COMPARISON, WHICH COMPARES LEVELS.

As with all primes, what the domains, levels and, behaviors actually are depends on the nature of the given. In our universe, for example, comparisons usually involve the seeking of stability or the lowest energy level. Raindrops, bubbles, plants and, stars tend to form roughly spherical shapes because it is the shape requiring the least energy. Radioactive decay of some heavier elements is for the purpose of seeking stability.

All the comparison prime does is compare levels, but this is the root of everything that has ever happened or ever will happen. As well as negative events, that which did not happen. No matter what the positive or negative event, it happened because a comparison decided that one level was higher than another. Everything that exists does so because a comparison decided that it should.

When we say that something is possible or not possible, we are saying that a comparison has been made between two levels and one or the other level has been determined to be higher. Of course, as with everything else in primes, what exactly those levels are depends on what given is manifesting the primes. In our universe, manifestation of the comparison prime is drastically affected by the fact that more than one object cannot occupy the same place at the same time. If there was another universe where 672 objects could occupy the same space at the same instant, the comparison prime would make some very different decisions but the primes would be no different.

We have seen how the domain prime has to come before the level prime but that the level prime can affect the existence of domains if the domain is composed of sub-domains. In the case of sub-domains comparison of levels, whatever the domains and appropriate levels may be, determines whether the larger domain will form or not.

All domains act as antennas for instructions from comparisons of levels that are always taking place. The comparisons compare levels and give the appropriate feedback to the domain. This happens whenever anything of any description exists. The only variable is the given that is manifesting these primes and the nature of this given.

In our given, our universe, all material domains undergo continuous comparison in each degree of freedom. The purpose of comparison in this case is to determine whether the domain should move or stay put. If move is decided upon, which direction and with how much force and if applicable, what is the dominant domain. The degree of freedom as with everything except the primes, is dependent on the given, the physical universe.

THE FOURTH PRIME: COMPENSATION

Once the first two primes; domain and level have been manifested, comparison comes into manifestation to continuously monitor levels and decide on appropriate action. This action, whatever it may be, taken as a result of a comparison, is called a compensation. Thus, we have our fourth and last prime.

POSTULATE 12: THE FOURTH PRIME IS COMPENSATION, WHICH ADJUSTS LEVELS TO SEEK STABILITY ON INSTRUCTION FROM COMPARISON.

Comparison and compensation go together like hand and glove. It does not make much sense to make any kind of comparison unless some kind of action will potentially result. It makes no sense to make a compensation without knowing what we are compensating for, which has been revealed by some kind of comparison. The comparison must come first and any action taken as a result of a comparison is called a compensation.

Levels are compared by the comparison prime and if some kind of change is called for as a result, we get a compensation. In our universe, any kind of change, any form of motion is the result of a comparison and compensation. Any kind of change involves some change in levels and an action may be thus required to compensate. Of course, no matter how many levels of sub-domains we have, the compensation will take place on the appropriate level.

POSTULATE 13: ALL CHANGES OF ANY KIND ARE MANIFESTATION OF COMPENSATION AND ARE EXPRESSIBLE IN LEVELS.

Generally, in our universe, comparisons not resulting in compensations usually result in stability, equilibrium or, continuation of the status quo, while compensation results in action. Stability can be defined as a lack of potential for compensation to be manifested, caused by stability in all relevant levels. When something new is added, inevitably as a result of the compensation prime being manifested, another compensation may be required to be manifested before all manifestations of comparison state that no further compensation is called for. Any compensation is a domain in itself whether it is a simple compensation such as a collision or a complex compensation such as the running of an engine.

POSTULATE 14: MANIFESTATION OF COMPENSATION IS A DOMAIN IN ITSELF. THIS CONNECTION FROM THE FOURTH PRIME BACK TO THE FIRST PRIME IS CALLED THE PRIME CYCLE.

A compensation could be negative as well as positive. If a compensation prevents something from happening rather than causing something to happen, we could call it a negative compensation. This is dependent on the given and makes no difference at all when considering the manifestation of primes.

Of all the four primes, it is obviously compensation which requires the most for it's manifestation. A compensation always overcomes a barrier, the compensation would not have happened except that some kind of levels change called for it. When we speak of something being either possible or impossible, we are discussing a compensation and are saying that one level either can or cannot be made to exceed another level.

POSTULATE 15: ALL COMPENSATIONS OVERCOME A BARRIER.

The issue of domain forming from sub-domains is an issue of compensation. Can the compensation raise the level of reasons for the domain forming above the level of reasons for the domain not forming? If so, the domain will form. If not, it will not form. Whatever the domain is as well as whatever the levels are is of course, a property of the given, whatever it is. The reasons for a domain not forming can be referred to as a barrier. In our universe, this barrier is often simply space or, distance. The making or breaking of a domain from or into sub-domains is always a compensation.

The formation of the given, what we call the big bang, was a compensation in itself. The primes existed before anything in the physical universe. When considering primes, we could say that the big bang was the "primary compensation".

As busy as the world around us, and the universe as a whole seems, compensation is actually exceedingly rare compared to comparison, at least above the atomic scale. In our universe, heat and radiation are the great melting pots of compensations. Any type of movement, transportation or, change from the explosion of the big bang to reflection and refraction of light is a manifestation of the fourth prime, compensation. Precipitation, osmosis, heat and, radioactivity are all compensations. Going to sleep when you feel tired, eating when hungry or, drinking when thirsty are just as much manifestations of the compensation prime.

Once we understand compensation and how it relates to the other three primes, it becomes obvious that, at least in our universe, a great cycle of compensations can get started. One compensation will change levels and cause another compensation, which will change levels and so, lead to another and so on indefinitely. Compensation being manifested will alter levels that will bring about other compensations. Some of these compensations in our universe involve the creation and destruction of domains.

As in all primes, actual manifestation depends on the given. In the case of our universe, the manifestation of the compensation prime would be utterly changed if multiple objects could occupy the same space at the same time. This is the most important factor of compensation manifestation in our universe. Collisions are not primes but are due to the nature of the given. There is only a given amount of space and time and only one entity of matter can occupy the same space-time. There is likewise only a given amount of matter and energy and this is why we often see levels tradeoffs as compensations.

This is what drives all the motion in our universe, what we could call the cycle of compensation. Isaac Newton wrote of actions and reactions but in the Theory of Primes, every action is already a reaction, a compensation. An action cannot come about any other way.

Newton's laws all concern manifestations of the compensation prime. The first law states that every action results in an equal and opposite reaction. This is a compensation in which one level is equal to the original action and one level is diametrically opposite to the original action. The original action of course, must have been a compensation in itself, which we could trace right back to the big bang, which is why we can refer to the big bang as the primary compensation.

The second law concerns momentum. Newton wrote that an object at rest tends to stay at rest and an object in motion tends to stay in motion until stopped by an outside force. This is a simple issue of comparison and compensation with the momentum expressible in levels.

The third law deals with gravity. Objects attract each other with a force directly proportional to their masses and inversely proportional to their distance apart. This is an issue of domains undergoing compensations expressible in levels.

CHAPTER THREE

▼

PRIMES IN ACTION

What exists and what happens depends on a combination of the primes and the nature and properties of the given. In order to become more familiar with primes, why not take a look at and analyze some more good examples of primes in manifestation.

I would also like to mention at this point that it seemed to me that the third and fourth primes, comparison and compensation, are separate primes. Some readers however, after seeing more of how primes are manifested in our universe, may form the opinion that the two should be considered as a single prime, leaving us with only three primes.

The earth is a domain. The surface of the earth can be divided into two sub-domains; land and water. These two sub-domains are the result of levels. The earth's surface is uneven (at different levels) and there is only a certain amount of water, a level of water.

Comparison and compensation take place. Depending on this level of water and the levels of the surface of the earth, some of the land ends up below the water surface when it is allowed to flow freely over the earth and, the remainder of the land remains above the water. The two familiar sub-domains of the earth's surface, land and water, are the result.

Anything that moves in our universe does so due to comparison and compensation. Any change or action is always a compensation and any compensation changes levels. A moving object is undergoing continuous comparison and compensation. When the comparison prime is manifested, it decides by measuring the appropriate levels, that the level of reason for the domain to move in the given direction is greater than the reasons not to move in the given direction. The result is compensation, in other words movement. The comparison keeps the domain moving in that direction until another level rises which calls for a different compensation. Such as a brick wall calling for a flying ball to go in a different direction by altering the existing levels relevant to the ball.

A compensation does not have to be motion, it can be any change in levels. It is just that our universe is spatially-oriented and motion friendly and this is where we often see compensation manifested.

In our universe, motion is usually in a straight line. But, when we walk or drive somewhere, we may not always go in a straight line. This is just as much due to comparison, however. Seeking the path of least resistance is the object of the comparison prime. As the compensation, we walk or drive in the direction we want to go along whichever route is easiest or most convenient. In our universe, momentum is often the deciding factor in choosing the path of least resistance. In other words, comparison choosing a particular compensation because it is a continuation of the previous compensation.

The same principle applies to water flowing through a channel or pipe. The water flows due to a comparison of energy levels and a resulting compensation. The energy levels are formed by the earth's gravity and water naturally flows from a higher altitude with it's higher kinetic energy level to a lower altitude with it's lower kinetic energy level. If the water does not follow a straight line in doing so, it is due to another comparison of the path of least resistance. If we were measuring electricity flowing along a wire, the manifestation of primes would be exactly the same.

Tuning a radio to a certain station involves a similar path of least resistance issue, varying a capacitor to make the radio most sensitive to waves of a certain frequency. The radio compares all signals of all frequencies and compensates by accepting the one with the highest level of resonance with it's circuitry. The generation of the radio waves at the transmitter is of course, a compensation for the currents pulsing at the given frequency in the transmitting antenna, which is of course a compensation for changes in levels in the transmitter circuitry. The entire radio issue is a manifestation of domain, the receiver joining the domain of the transmitter as the desired station is tuned in.

Soldiers marching on a parade ground is also an issue of domain and sub-domain. If twenty new recruits are just standing around with long hair in civilian clothes, we could say that we have twenty domains. But if they are all in uniform and marching in step, we definitely only have one domain. Since domain and sub-domain is always a manifestation of levels when we view it as one domain, we are seeing it from a higher level than if we view it as twenty domains.

An interesting property of levels concerning sub-domains is that levels are not analog, or existing over a gradual, sliding range. But rather are quantized, or jumping straight from one level to another with a step rather than a gradual transition. In this example, for instance, we can see either twenty domains on one level or one domain composed of twenty sub-domains on the next level. There is no in-between or gradual transition. When dealing with liquid or gases, the sub-domain is as small as the component atoms or molecules. When dealing with solids, the smallest domains of the solid become the sub-domains.

The earth is a domain speeding through space in a straight line. Another domain comes into play, which we know as the sun. There is a much larger domain associated with the sun, it's gravitational domain. The earth is within the sun's gravitational domain. A comparison results, the level of the sun's pull on the earth and the level of it's momentum through space. A compensation results from this comparison, the earth will not continue on it's straight-line path through space, nor will it be pulled into the sun. The earth will instead accomodate both forces by going into an orbit around the sun. This is all a matter of the given, not of the primes.

The earth, the sun, land and, water is made up of atoms. An atom is a domain. Electrons orbit the central nucleus in much the same way that the earth orbits the sun. What kind of atom forms depends on a level, the number of protons in the nucleus. The periodic table is levels of the same domain, the atom. The hundred or so different atoms can combine into millions of different molecules, which is a manifestation of domain and sub-domain.

The number of electrons orbiting the nucleus in an atom is normally at the same level as the number of protons in the nucleus, resulting in a neutrally charged atom. When these levels are not equal, we have what is known as an ion, which is an issue of domain and sub-domain. The charge on an ion is the result of a compensation caused by comparison determining that the levels of positive protons and negative electrons are unequal. When we vary the number of neutrons in the nucleus of an atom, we are also dealing with domain, sub-domain and, levels. This is what we call isotopes.

The existence of atoms leads to chemical reactions. Some atoms combine with each other and some do not. Atoms combining is an issue of domain and sub-domain. Whether the atoms combine or not is an issue of comparison and compensation. Whether the reaction of atoms combining is exothermic, or gives off heat as a compensation, is an issue of comparison of the normal energy level the atoms hold when separate and then when together.

Electromagnetic radiation such as light, is the result of comparison of levels and compensation. This is the most common compensation by far in our universe. When electrons in a higher orbit in an atom fall to a lower level orbit, the electrons require a lower energy level. This disruption in the existing levels results in excess energy leading to a compensation in the form of light or other electromagnetic radiation.

A river flows by seeking the lowest energy level by compensation following comparison. The water, like any other domain, is in a continuous state of comparison of levels to seek the path of least resistance. A dam built across the river would be the ideal example of a barrier, something preventing a compensation that would have happened if the barrier had not been there.

A machine of any description is naturally a domain in itself. Whenever we have any type of motion in our universe, it is always a compensation as a result of a comparison. Efficiency and coordination are levels. Any machine that I can imagine operates by enabling the variation of appropriate levels in order to lead comparison to produce the sought-after compensation.

Arithmetical operations such as plus and minus are issues of domain and especially the size (or level) of that domain. Multiplying, dividing and, exponents are issues of sub-domains. The difference between an equation and an inequation is a matter of comparison of levels. Geometry always deals with domains, the measurements of these domains are levels. Whether two lines are parallel or perpendicular is a matter of levels. Trigonometry deals with levels, the three angles of a triangle always add up to 180 degrees. Trigonometric functions are domains composed of ratios of levels. When we try to find an optimum point using the highest point of a polynomial with calculus, we are concerned with levels. Analytic geometry describes the location of a point with two levels, the level of X and the level of Y if we are dealing with two-dimensional geometry. Three levels, X,Y and, Z if we are dealing with three-dimensional geometry.

If we apply the primes to the English language, we find that most question words are domain words and include this, that, them, those, the, a, an, he, him, his, she, her, hers, this and, that. Level and comparison words include when, who, what, where and, there. Compensation words would be why and how. A noun, which is a person, place or, thing, is always a domain. A verb, an action, is always a compensation.

PATTERNS: COMPLEX PRIMES

No matter what exists, no matter what happens, it is always a manifestation of the four primes. The examples that I have given above are situations that tend to neatly display manifestation of the primes. However, our universe and our world is very complex. Things that we see often do not appear to fit neatly into my

Theory of Primes. This is simply because primes can be manifested on an infinity of levels, limited only by the given, which in our universe is space, time, forces, matter and, energy.

Aside from primes, we can see other patterns that occur often that I will call simply that: "patterns". We could also call patterns "complex primes". In the same way that relatively few atoms can combine into millions of different molecules, primes can combine into patterns. Just as most of the colors that we see in the world around us are not the three primary colors, but complex combinations and shades thereof. Understanding these patterns would make the primes much easier to understand by enabling us to keep primes and patterns separate. In fact, the Theory of Primes could probably also be called the "Theory of Patterns". Just remember that everything is manifestation and combinations of the basic primes.

Before going any further, there are two things that must be understood. 1) All patterns consist of and can be broken down into primes. 2) Patterns are dependent on the given for their very existence while primes are dependent on the given only for their manifestation.

In other words, patterns are only manifested when the given makes it possible. Primes are manifested whenever there is any type of given. The patterns result from the nature of the given while the primes do not. Once again, whether we see the complexity of patterns or the simplicity of primes depends on what level we are looking on. The only reason for discussing patterns in the Theory of Primes is for understanding. It is often in patterns, rather than in the basic primes, that we begin to see clearly our familiar world.

So far in this book, I have repeatedly used the term "In our universe". One easy way to separate primes from patterns is to imagine another universe where everything about the given may be utterly different from our own. It is doubtful if we would find the same patterns being manifested in such a universe as in our universe but we could be absolutely sure that we would find the same primes being manifested. Moving on from our discussion of fundamental primes to our familiar environment requires more understanding of the nature of the given in order to comprehend the patterns it produces.

We have already met two of the common patterns found in our universe; boundaries and, barriers. Let's have a look at some more in order to be better able to keep them separate from the primes. Just as it is necessary to separate the primary colors from all the shades and pastels if we are to understand color, or to separate molecules from the constituent atoms if we are to understand atoms.

If a barrier creates levels by making one domain into two, then a bridge does the opposite by making two domains into one. A bridge is anything that enables a barrier to be crossed. A gate in a fence would be a bridge as would a language translator. Soap is also a bridge, dirt and water often stay separate but a molecule of soap has one end that is attracted to water and another end that bonds with dirt. This enables water to carry away much more dirt than it would without the soap.

We have also examined another pattern, the path, which is the route of least resistance or most gain that a compensation follows. Path forms by continuous comparison but operates by compensation. A path makes two or more domains into one as does a bridge, the difference being that path does not involve any barrier, while bridge does. Of course, like anything else, path and bridge are domains in themselves. Domains existing especially for compensations to be manifested in. A most obvious example of the path pattern would be the orbits of the planets. Another would be the arteries found in biological life.

A pattern that we have not yet discussed is seed. A seed is a domain that results in the creation or destruction of another domain. Any compensation is potentially a seed but a complex seed initiates a comparison-compensation cycle in the given which brings about a change of some kind. Anything in our universe that is self-perpetuating or starts a chain reaction is a seed.

Commonly seen seeds are growth poles such as a village which becomes a city or a snowball rolling down a hill and getting larger in the process by gathering more snow. Condensation nuclei upon which cloud droplets form in our atmosphere are seeds. The seeds which grow crops are, of course, seeds. A flame in a dry haystack is an example of a seed. A neutron in a nuclear chain reaction is a seed. The growth of an idea is a seed pattern.

We should differentiate between a seed and a catalyst. A seed starts a compensation cycle but is not necessary to make it possible. A catalyst is a kind of a cross between a seed and a bridge in that it is necessary to make the change possible. One of the best examples of a catalyst is an enzyme involved in a biochemical reaction. A matchmaker who introduces a couple who otherwise would not have met is another good example. We could say that potatoes were one of the catalysts of the industrial revolution.

A pattern that seems to be manifested all over our universe is series. Domains that come into manifestation as variations of previous domains. Series is an accomodation of change and momentum. Anything that is repetitive in nature and the patterns of the given is a series. Motion of an object through space forms a series of events in space-time. When a star explodes and other stars form from the debris, we have series being manifested. when a living thing dies and the component proteins and nutrients enter the ecosystem and form a part of another living thing, we have a series. Generations of living things is one of the most obvious series. The series pattern is closely related to path and is a cross between path and alphabet or sub-domain.

Whenever we consider domains and sub-domains in our universe, we inevitably encounter alphabet, another common pattern. Alphabets consist of a series of domains, especially suited to serving as sub-domains due to the complementarity of the member domains of the alphabet. The only differences between alphabets and sub-domains is the level of complexity, the level of order and organization and, the frequency of occurrence. Complementarity is vital to an alphabet and can be defined as simply the tendency for sub-domains to form domains.

I also think that potential of separate existence is important in an alphabet. For example, I consider the human liver to be a sub-domain rather than an alphabet member of the human body simply because the liver usually does not occur outside the body. Primary purpose is also a vital difference between alphabet and sub-domain. The primary purpose of letters is to form words and the primary purpose of amino acids is to form proteins. So, I consider letters and amino acids as more alphabets than sub-domains.

If there is no order that sub-domains must go together to form a domain, then the sub-domains do not constitute an alphabet. For example, a chunk of limestone under Niagara Falls is a sub-domain of planet earth and not any kind of alphabet member because not only is it not essential for the earth to be a domain at all but it could be just as much a part of planet earth if it was at the bottom of the Indian Ocean.

Complementarity and thus alphabets is entirely an issue of the given. Commonly seen alphabets are elements forming compounds, letters forming words and, amino acids forming proteins. Complementarity can best be seen in chemistry, such as acid-base and oxidation-reduction reactions, and in the negative electron-positive proton attraction. Of course, we usually spend more time thinking about the male-female complementary attraction than the other two examples for some reason.

The pendulum is a common pattern seen on earth, especially with regard to living things. The supply and demand swings of capitalism and, the populations of predator and prey are commonly seen manifestations of the pendulum pattern. In physics, the pendulum pattern is usually referred to as oscillation or simple harmonic motion.

The pyramid is one more pattern seen throughout the universe. The essence of the pendulum is a few at the top and many at the bottom. The pyramid pattern applies to the food chain, the formation of large and small stars, rich capitalists and laborers and, leaders and subjects.

THE PATHWAYS OF OUR UNIVERSE

To get a better perspective on the primes and patterns, let's take our viewpoint up yet another level to the operation of the universe as a whole. We will look at the universe primarily from the time perspective and I will call the day to day operation of the universe "the time pathway".

Every compensation in our universe, every action, has both a time and a space component. Comparisons are instantaneous but due to the nature of our given, compensations take time, especially if many compensations in the sub-domains of a domain are required. It is this series of compensations, the compensation cycle, which drives our universe in the time sense.

This is what gives us time. The space-time that originated with the big bang included time as a fundamental component of the given. However, it would have no meaning if compensations in our given took place as instantaneously as comparisons. If this were the case, time would still exist theoretically but everything would already have happened instantaneously. Time would be meaningless because there would be no motion at all beyond the one instant.

We could say that time is manifested by motion, if nothing is moving then time does not effectively exist. The units that we define to measure time, such as day and year, are always based on motion, of the earth in this case. I believe that God created the rule that nothing could travel faster than the speed of light to prevent this. However, the speed of light, or anything else to do with time, is an issue of the given and has nothing to do with the primes. It is delay that makes the universe operate. If every compensation was instantaneous, everything would have already happened. There would be space but no real time.

Likewise, time is manifested by motion and in a condition of absolute zero, there is no movement and hence, no real time. In a black hole, time and space as we know it cannot exist, but the primes would be unchanged. A photograph is all we need to see that time is a part of the given and does not affect the primes. There is no movement in the photograph and hence, time is non-existent. Yet domains and levels are just as much in manifestation as anywhere. Domains manifested in whatever is in the photograph and levels manifested in colors and shades in the photograph.

The issue of motion in the three space dimensions involves levels and compensation from a primes-only point of view. Pathways, direction and, velocity are levels and are all properties of the given. There is no real direction in the universe as a whole, we define north and south only by the orientation of our planet and if we go down to the level of the primes, direction has no meaning whatsoever.

Anything able to move in our universe has both a space and a time component but has freedom only in the space component simply because there are three dimensions of space but only one of time. Movement is possible in two directions in each of the three spatial dimensions but only in one direction in the one time dimension. There are infinite possibilities for motion in space but only one choice as far as time. We call the different dimensional motion that movement can take "degrees of freedom".

However, when we look at what the universe is doing as a whole, we look primarily at time rather than space. Everything that happens in our universe is part of a tree structure of compensation cycles that can be traced back to the big bang, at the root. The big bang is the ultimate compensation and since compensation is the last of the primes, the primes must have existed before the big bang.

What about when something happens, in other words an "event". What we call an event in our universe, from an exploding star to a birthday party, is simply a domain in the time dimension. We could say that events are domains of primes and patterns. Just as any matter is an arrangement of atoms, anything that happens, an event, is an arrangement of primes and possibly patterns. Just as domains can consist of sub-domains, events consist of sub-domains and possibly sub-events.

A party is just as much manifestation of domain as an asteroid or a planet. Events are domains of primes and patterns. The dividing line between domain and event is that event involves compensation while domain does not, at least not in the dominant domain. Both involve space and time, domains are more space-centered, events are more time-centered.

Not only do we have the fundamental primes, our universe contains the patterns such as those we have seen. This makes a flexible alphabet of primes and patterns. What can happen with such an alphabet is limited only by the nature and properties of the given.

Of course, the differences between domain and event has meaning only in our given, our universe, if we think only of primes, it has no meaning whatsoever. An event is a domain and no domain is any more of a manifestation of the domain prime than any other domain.

One thing that is absolutely vital to understanding primes is to remember that we see our familiar world with our own lenses in ways that are relevant to our daily lives. In the Theory of Primes, there are no events, elements, molecules or, reactions. There are only the primes in manifestation. The advent of the given, the universe, had to follow manifestation of the primes just like anything else.

Primes are at the deepest level of physical existence, which is why they are called primes. When preparing dinner, we could say that we are cooking. We could go deeper and describe what the atoms and heat energy is doing. We could go deeper still and describe the manifestation of the primes. Terms such as cooking, chemistry, physics and, astronomy exist only for our convenience as we see with our own lenses.

If we could step out from our daily lives and see the infinite as well as the infinitesimal, the instantaneous as well as the eternal, I believe that we would see the primes with much more clarity. It is doubtful that it would have taken until now for someone to point out the primes.

CHAPTER FOUR

▼

THE LIVING DOMAIN

A NEW LEVEL OF REALITY

As proof of my Theory of Primes, or at least strong evidence of it's veracity, we are going to look at a higher level of reality altogether and see if the same four primes are manifested. Thus far, when we have considered the given, our universe, we have for the most part looked at matter, energy, forces and dimensions. In other words, inanimate matter.

My case in favor of this Theory of Primes is that human affairs, which most people would agree is a completely different realm than the physics and chemistry that we have referred to thus far, nevertheless manifests exactly the same primes. We do not have a different universe that we can go to in order to see if the same primes are manifested. What we do have however, is completely different fields of knowledge. Examining primes in the context of human affairs will also give the added bonus of making the Theory of Primes easier to understand.

All fields of knowledge from physics, chemistry and, astronomy to geography, geology, history, computer science and, politics manifest exactly the same four primes. Which is of course, what we should expect if the primes are as primary as my Theory of Primes is claiming.

We will refer to the level of inanimate matter as the "inanimate level" and consciousness as the "conscious levels". Plural because there is different levels of consciousness.

Human affairs is a higher and different level of reality than the level of inanimate matter. Reality is a domain and it manifests levels like other domains. All inanimate matter is at the lowest level of physical reality. Living things, beginning with protozoa, use the patterns of seed and series to replicate, this is the beginning of higher levels of reality.

Human beings, as far as we know, are at the highest level of physical reality, meaning that we are on a higher level than a rock but are still subject to the same laws of physics. The spiritual realm is at a higher level still that is beyond the material universe and so is not subject to the laws of physics. The scientific tools that we use are made of and therefore are on the same level of reality as inanimate matter. This is why telescopes are useful in exploring the universe but are of no use in solving human social problems.

The higher level of reality of human beings means that we are subject to a wider range of reality than a boulder. A boulder is just as much a manifestation of the domain prime as a human being and manifests levels, comparison and, compensation. The difference between a human and a boulder is that the boulder does not manifest thoughts, visions and, feelings in the domain and level prime as humans do. Boulders do not have senses such as sight and hearing to affect the comparison prime as the human mind does.

If we define reality as anything that is real or anything that affects anything that is real, it becomes easier to see the similarities as well as the differences in the reality levels of a boulder and a human being. A boulder is affected by gravity, heat and, physical impacts as humans are. A boulder is not directly affected by myths, optical illusions or, dreams. Although someone can envision the building of a road or garden and move a boulder to do so.

In higher levels of reality, knowledge of reality, or consciousness, becomes an important part of reality in itself. A boulder has no knowledge of reality to affect it's comparison prime while human beings do. To a boulder, something must be physically real to have reality. A boulder hurtling through space cannot collide with a dream.

To a boulder, memory of a long-dead person, a rainbow, knowledge of a distant country, a song, a beautiful blue sky, an inspiring speech or, a mythical legend have no physical reality and have absolutely no affect on it's primes. To human beings, all of these things can be very real and can have a profound affect on the manifestation of the primes. This reality is possible because humans exist at a higher level of reality. Dreams cannot exist at the level of inanimate matter but thrive at the level of human beings.

Any escalation to a higher level usually involves a tradeoff of some kind. A human being enjoys the advantage of range of reality over a boulder resulting from the higher level of reality of the human being. The tradeoff is that there is a down side to this wider range of reality. A human being can be affected by mistakes, lies or, be taken by surprise while a boulder cannot. A boulder cannot enjoy the sight of a beautiful garden but neither can it make mistakes.

The primes are just as easily seen manifested in human affairs as in physics and astronomy. One set of opposing levels that are often manifested from the domain of a human being is possesions vs money. An individual's possessions or money are domains which can be expressed in levels. The levels can change relative to each other by a compensation in the form of a purchase or sale. Such a comparison must be preceded by a comparison; is this worth selling or buying for this price? Value, meaning that the addition of a domain is more desirable than it's subtraction, is a manifestation of level.

Any human being has two sub-domains, that of desires and that of possessions. Quality of life is a level and can be defined as the level of coordination between these two sub-domains. Any type of exchange done by a human being, such as sale or purchase, is a compensation done as the result of a comparison.

Knowledge is another domain that can be expressed in levels. A boulder has no knowledge. Maturity is a levels issue as is health. Life and death are domains. A human body is a domain with bodyweight, height, age, blood pressure and, muscle to fat ratio all manifested as levels. Freedom is a domain that manifests levels. Danger and safety is an issue of stability, which we have defined as the lack of potential for compensation, which is a manifestation of level.

Exactly the same primes are manifested in the lives of animals, fish, birds and, insects as in the lives of humans. Species is a complex manifestation of sub-domain. Birth is an issue of a sub-domain reaching a certain level, at which point it becomes a separate domain. Digestion is a matter of domain. The immune system is concerned with domain and works by comparison and compensation. Osmosis functions by a comparison of levels of domains, which may result in compensation. A cell is a sub-domain of the organ, the nucleus is a sub-domain of the cell.

Organic chemistry and biochemistry are entirely different subjects from inorganic chemistry but manifests exactly the same primes. Everything to do with living organisms from osmosis to the life cycles to increase of muscle mass by exercise is a manifestation of the four basic primes.

The mind of a human being is just as much a manifestation of the domain prime as a boulder, except with many more relevant sub-domains expressible in levels. Personality is a complex manifestation of all the basic primes. Talent and skill is a sub-domain as a sum, divisible into individual sub-domains manifested in levels. Human and animal habits are very much pathway and momentum issues, the comparison prime favoring a certain course of action because that is the way it has been done previously.

The priorities, goals and, focus of a human being are manifestations of domain and level brought about by compensation. Success and failure, like all opposites, are manifestation of domain but can easily be expressed in levels. Whatever the reasons for an individual's ambitions, such reasons are inevitably manifestations of the compensation prime. Perspective has so many possible sub-domains, all expressible in levels, that it would be most convenient to express it as levels on a grid as in describing the location of a point in analytical geometry. Political persuasion should involve less manifestation of levels as overall life perspective and would probably be expressed in levels from far left to far right. Those people with certain views manifest domains.

A person's interests are what we could call double domains, which are often seen when consciousness is involved. When a person gains interest in soccer, the person is added to the domain of those interested in soccer, while soccer is added to the domain of the person's interests. Gaining or losing interest is like gaining or losing anything else, a manifestation of all four primes, ending in a compensation.

Likes and dislikes are domains, manifested in levels as degrees. The desirability of something or someone is a level. Deciding if something is worth doing or a place is worth going to is a comparison, actually going there is a compensation. Adding something to your experience as well as adding something to your possession is an issue of domain.

Patterns, or complex primes, that we have discussed are also very visible in manifestations related to human beings. For example, the tendency for son to follow father would be a manifestation of the series pattern. The tendency for boys to lean toward certain tasks and girls toward others is a manifestation of the alphabet pattern.

Like so many other things relating to humans, friendship is an issue of domain that is easily expressible in levels, just as so many inanimate domains are easily expressible in levels. The pattern of complementarity also comes heavily into play in the issue of friendship. Change of mind is an issue of domain, brought about by compensation. Feelings are levels which can be grouped by domain.

A question in a human mind is a comparison manifestation. Thoughts, memories and, nostalgia are all domain issues, like any human feelings are expressible in levels. The special places, special people and, focal points of our lives are also issues of domain expressible in levels. Art, music and, beauty are manifestations of the complementarity pattern and therefore the domain prime, once again expressible in levels.

HUMAN AFFAIRS

The given fabric of the universe could have been completely different and the primes would have been the same. This is shown by the fact that exactly the same primes govern human affairs. No matter how different this universe or another universe could be, the primes would be the same. The primes are manifested in exactly the same way whether the given is atoms, sports or, politics. I consider the fact that human affairs manifest exactly the same four primes as sub-atomic physics as sufficient proof of the veracity of the Theory of Primes.

The entirely different given and situations at the higher levels of reality makes not the slightest bit of difference in the manifestation of primes. You can go into a library and read a book of any description-fiction, geography, politics, business, history or, science and you will see exactly the same primes in manifestation.

What about the machines which human beings design and build? Simple tools, which are domains in themselves, operate by an issue of levels and compensation. Machines are, in fact, domains that use compensation in order to achieve levels trade-off. An elevator decreases the energy level of it's power source in order to raise it's payload to a higher energy level from the point of view of the kinetic energy of gravity. When considering primes, it makes no difference whatsoever whether the given is "natural", such as the flow of a river or, "synthetic", such as a machine. Exactly the same primes are manifested.

We speak of the "efficiency" of a machine. A bicycle has a very high efficiency, a steam engine has a very low efficiency. This concept of efficiency is an issue of domain. Does the entire compensation fall within the domain for which the machine was designed? If not the machine, like most machines, are less than 100% efficient. The efficiency can be easily expressed as a manifestation of the level prime.

Transportation is an ancient issue of machines that is a very good illustration. If we are moving something northward, we willingly decrease the level of our "southwardness" in order to increase the level of "northwardness" of our passengers or pay-load. As with any machine, it's use is in itself a compensation. Before beginning the journey, we decided by a manifestation of the comparison prime that it was more desirable to undertake the journey, which is a manifestation of the comparison prime, than not to undertake it. In addition, the comparison that prompted us to send our vehicle northward also "won" over the infinity of simultaneous comparisons prompting us to send the vehicle south or west or east and so on for all points of the compass.

When we consider the entire economic system, exactly the same primes are found. A stratified class system of commoners and elites is clearly a levels issue, expressible in domain. Capitalism is a continuous display of all the primes in action from the domains of shops and the sub-domains of goods to the levels of prices to the comparison of purchase and manufacturing decisions to the compensation of raising prices, hiring new workers. Supply and demand is a continuous kaleidoscope flow of changing levels, comparison and compensation.

All issues of prices and values are primarily concerned with levels. Deciding whether to buy something is a comparison. Actually buying it or not buying it because the price is too high is a compensation. We should point out that deciding not to buy something is just as much a compensation as buying it. A negative compensation, creating a negative event, is just as much a manifestation of the compensation prime. Quality of goods is a manifestation of level. Whether different customers may have different definitions of quality is irrelevant to the manifestation of primes.

Money itself is an issue of domain. The money that the government issues is a domain equal (which is a level) to the value of all goods in the economy. When there is more money or less money in circulation, the comparison prime manifests the compensation prime by lowering or raising the value of each unit of currency (dollars, pounds, francs, gold, etc.)

The use of currency is much more convenient than simple barter due to manifestation of sub-domain. A barter transaction cannot be done unless there is something of the value of the desired item to exchange or to exchange for. It is for this reason that bartering would be a very awkward system if implemented in a modern economy. With currency, in contrast, no such limitations exist. The money that is circulated forms a domain that is especially easy to divide into sub-domains down to the lowest unit of currency, which many nations call a penny.

Money or currency effectively creates a parallel or shadow domain to that of the goods in the economy. However this also introduces another level, that of confidence in the financial system. When this confidence is at an adequate level, all is well. But when it is not for some reason, a manifestation of the compensation prime known as a run on the banks may be seen. The health of the economy, which is a manifestation of the level prime, decreases below that of what it would be in a simple barter economy.

In a capitalist economy, the speculation of currency, bonds and stocks is a mirror of the rest of the capitalist system in it's complex and ever-changing levels (prices), comparison (buy and sell decisions) and, compensation (purchases and sales).

The basic human enterprises such as plant and animal domestication is an issue of domain. Making wild plants and animals into a sub-domain of the larger human domain. An industry practiced by humans such as freight transportation is a domain, each individual business dealing in freight transportation is a sub-domain. Resources are a domain, industry operates by comparison-compensation to create a finished product, which is also a domain.

A corporation is a sub-domain of the industry that it partici-
pates in. An engine block manufacturing facility is a sub-domain
of the auto industry. Each department or office of that corporation is a sub-domain of the corporation. Each individual is a
lower level of sub-domain in the corporation. Any business
enterprise is a domain formed by compensation following a
comparison. In other words, someone noticed a chance to make
money and started an appropriate business in order to do so.

A job is a sub-domain both of the company and of the
worker's life. In patterns terminology, we could say that jobs are
bridges between the company and it's employees. Authority is an
issue of domain as well as level. Authority covers a certain
domain, the larger the domain covered by the authority, the
higher the level of the authority. While a job is an issue of
domain, a career is more a manifestation of the series pattern.

Since jobs and careers is something that most of us are very
familiar with, this is a very good illustration of the independence
of the primes from the given. When considering primes and patterns, all jobs and all careers are exactly the same. From a sugar
cane harvester in Cuba to a drug dealer in any city to the CEO of
a multi-national corporation in London, the only difference is in
the given.

All economic and political systems from the hierarchies and
roles seen in the social orders of many animals to modern capitalism and democracy are domains. All operate through manifestations of levels and compensations. The primary difference in
economic and political systems tends to be in the comparison
prime.

Democracy for example, often consists of two parties. Each of
these parties is a domain. The pool of registered voters is also a
domain. In democracy, the passage of time is divided into sub-
domains by events, which are domains, acting as boundaries

known as elections. An election is an ideal example of a manifestation of the comparison prime. The infinity of comparisons that take place every second all across the universe can all be thought of as "elections". At an election, a comparison of levels is taken. The level of the number of voters preferring one party compared with the level of the number of voters preferring the other party. If the outcome of the election is that the incumbent party gets the most votes, no change occurs. In terms of primes, we would say that no compensation takes place. However, if the party that is not in power is decided by the comparison to have gotten the most votes, a compensation does take place. The opposition party takes over as the ruling party.

It should be easy to see that when breaking democracy down into it's component primes, exactly the same primes are manifested as when dealing with atomic physics.

This is significant evidence in favor of the Theory of Primes.

The legal system in any society is a domain manifesting all the other primes. The domain of laws functions by manifesting the comparison prime. In the jury system, a comparison is made between levels of evidence for and against to decide whether the defendant has broken the law. If the decision is "guilty", then a compensation takes place and the defendant is punished. The level prime is once again manifested as the severity of punishment.

The same primes are manifested in any "legal system" from the courts to school and workplace reprimands. The law itself can exist on different levels. For example, if in the workplace you were close friends with your supervisor and a central member of the clique, you may get the benefit of a different level of considering reprimands than someone who was not.

A war manifests the same primes as the legal system. The war itself is a domain consisting of two sub-domains. Each battle is a comparison of levels between the two sub-domains, which are the two sides. Whichever sub-domain the comparison favors manifests the compensation prime by advancing following the battle while the other side manifests the compensation prime by retreating. One side eventually does a comparison and determines that the war effort is not worth the cost and destruction. This side then compensates by surrendering or offering negotiation.

A modern war usually has sub-domains manifested as land battles, naval battles and, air battles. A war can be considered from different levels. An infantryman, a leiutenant, a general and the president or prime minister all see the same domain from different levels.

Culture, which means way of life, and which every human society has in some form, is a manifestation of the same primes. Culture is a domain with levels of many different things, formed over time by the comparison and compensation of everyday life.

A nation is a domain which forms by compensation of sub-domains. Ethnic groups are sub-domains of the national domain. An empire is a part of the domain on a different level than the nation proper.

The things which are of significance, importance and, of value are levels which in manifestation comprise a culture. Significance is a level that could be expressed by domain. A warm coat has a high level of value in Yukon but a low level of value in Mauritius. A woman with bare breasts picking crops in certain domains would get no unusual reaction. The fourth of July has a high level of significance in a domain known as the United States but, no extraordinary significance in other domains.

Celebrations of immigrant groups are manifestations of the domain prime in the form of sub-domains. Those at the celebration are announcing that they are a sub-domain of the old country even though they are within the domain boundaries of the new country. Since the old country is in a different geographical domain, such celebrations are manifestations of the bridge pattern, a bridge between the celebration in the new country and the old country. Immigration is of course, a compensation brought about by comparison between the advantages and disadvantages of the old land against that of the new land.

Each human being is a domain with a certain amount of knowledge, which is a sub-domain of the person that can be expressed in levels. The sum of human knowledge is a domain and the technical capability is also a domain. The knowledge and skills of each individual is a sub-domain of this total sum. The knowledge of each human being consists of a domain of levels. Each of us has a level of knowledge on topics ranging from the geography of our hometown and the names of people from school to the meteorological processes on Saturn and the growth rate of various crops in the soil and climate of western Mongolia.

Human consciousness is a domain on a higher level of reality than inanimate matter and this brings other domains into existence. To a human, there are viewpoint domains and defined domains, which would have no meaning if there existed only inanimate matter. If you look in a certain direction, literally or figuratively, what you see is a viewpoint domain. There would be no reason to consider this a domain if you, or any other human, did not exist.

A defined domain could possibly mean anything that humans have made which was not occurring naturally, such as a building, national borders or, plots of land. What we are going to consider as defined domains here is primarily units. Meters, yards, miles and, inches are defined domains. These units measure nature but would have no reason to exist if people did not exist. We could also say that names, such as rock and tree are defined domains. Colors are interesting as defined domains in that colors are merely creations of the human eye and have no real existence in inanimate reality.

Taste and preferences in anything from food to art is an issue of sub-domain that can be expressed in levels. The pattern of complementarity comes into manifestation here. If the food, art, etc. is desirable, then it will fit better as a sub-domain of the domain that we manifest as human beings. The concept of eating a balanced diet manifests the complementarity pattern so much that it brings the alphabet pattern into manifestation.

Literature and stories is just as much a manifestation of all the four primes as real-life. For that matter, it is just as much a manifestation of the primes as sub-atomic physics or the development of galaxies. A story is a domain just as the book or video that manifest the story is a domain.

Sports manifests all the primes. Each team is a domain just as each player is a sub-domain. The game itself is a domain. The scores are levels. Each goal or home run is a compensation. The action in any sport is continuous manifestation of comparison and compensation. The fame of an athlete or any other celebrity is a level. A football play is a domain.

THE SPIRITUAL DOMAIN

The four primes are so primary that we find the same manifestations of primes in the spiritual domain as the domains of human affairs and inanimate matter. I consider the spiritual domain to be the highest level of reality with human affairs and matter on lower levels. The fact that Jesus often used parables to make a point or provide a lesson shows that the spiritual domain manifests the same patterns and therefore the same primes. Ecclesiastes, in the Old Testament, wrote that "There is nothing new under the sun", meaning that the same human situations, the same patterns and, the same primes are always manifested no matter where we go.

The primes are morally neutral, just as there can be no such thing as a mistake in the manifestation of primes, there can be no good or evil in the primes in themselves. Good and evil is a given just as space-time, matter, forces and, energy is a given. Good and evil are domains that can be expressed in levels. An action, which is a compensation, can be in the domain of good or the domain of evil and can be described as a certain level in the appropriate domain. Murder is at a higher level in the domain of evil than stealing a chocolate bar from a vending machine.

For an action to be good or evil means that we must make it of our own free will. An action is a compensation that requires the manifestation of the comparison prime first. In the case of good or evil, comparison is the free choice in manifestation. Any influences in the choice of good or evil are of course domains manifesting levels which affect our compensation prime in it's manifestation.

Opposing situations such as true and false or negative and positive are also domain and level issues that require higher levels of reality for manifestation in the same way as good and evil. Except for electrical charges, negative and positive requires the higher level of reality in the form of human beings for manifestation. True and false also requires higher levels of reality for manifestation, human beings can be right or wrong but there is no such right and wrong or true and false when there is only matter in motion.

CHAPTER FIVE

▼

THE HIDDEN PRIMES

The primes must exist. I can see no logical way to doubt this theory that only four basic things can exist or can happen. We observe things, we put numbers on them. But, we have not noticed that primes have already defined the four things which can exist. Elements and forces in the universe have been categorized but the primes have been ignored. This is a classic 'can't see the forest for the trees' issue. The primes have not been so much hidden as camoflaged.

The more we think about it, the more it seems that there must be some type of rules for what can exist that must have been there before anything could exist. The universe is just too orderly for it to be otherwise. Try to imagine our universe without the primes to define what can exist.

My feeling is that primes fall more in the domain of mathematics than science. I cannot help wondering whether anyone has thought about primes before this. Surely, someone must have wondered whether there is basic rules for what can exist or fundamental patterns that everything must follow. Rules even more basic than the laws of physics.

The primes are a branch of mathematics more primal than even the counting numbers. The development of mathematics since ancient times continued from the counting numbers to more and more specific mathematics. The ancients had more use for counting and calculation than for searching for the primes. The primes were missed until now.

REASONS FOR ESCAPING ATTENTION

I am sure that someone before me must have thought of something along the lines of the primes. Primes are deeper than anything in the physical universe but not as deep as the spiritual realm and this is what caused the primes to escape attention. We tend to look at either the spiritual or the physical without considering the gap between them. Many people have tried to bridge the gap but apparently, no one thought to look in this gap. It seemed to me that the five great roadblocks to recognition of the primes are the lack of a control group, the direction of learning, the given, patterns and, the question words.

NOWHERE WITHOUT PRIMES

The primes are simply so obvious that we do not notice. There is absolutely nowhere without primes. This is one issue with a complete lack of a control group. We do not notice primes because there is nowhere without primes. The primes are everywhere, nowhere can possibly exist where there is no primes or where the primes are different from the four primes that I have described.

There is no such thing as a control group without primes and for this reason, perspective and experimentation with primes is impossible. We have other cultures to show us what it is like outside our own culture. We have animals to show us what it is like not to be human. We have the knowledge gained from moon landings to show us what it is like outside earth. We have nothing and never can have anything to show us what it is like without primes.

Imagine a species of fish that lives in the middle of an ocean. The fish spends it's entire life well below the surface but well above the ocean floor. The fish knows nothing but water. When we contemplate primes it is a lot like the fish contemplating water. Since the fish knows absolutely nothing but water, it will have great difficulty understanding the true role of water in reality.

The most distant and exotic places in our universe work with exactly the same primes that we do in our daily lives. Any spiritual realm that humans believe in, as well as the most far-out sci-fi story operates on the same primes that are manifested in a volleyball game. The same primes that are manifested in all of existence.

If you strain your mental powers and imagine a universe utterly different from our own, the patterns that form in our universe from combinations of primes may not be apparent but the four basic primes always will. It could be a universe without space and time or one with an infinite number of space dimensions and 11,327 time dimensions. The same four primes would always be manifested. The patterns manifested in our universe from combinations of the primes may be completely different but the primes would remain the same.

DIRECTION OF LEARNING

The primes have been left behind because the realm of primes is at the opposite end of the scale from the cutting edge of progress in science and mathematics. The primes are more primal than the counting numbers but were probably not noticed in antiquity because of the lack of apparent practical application at the time. People in those days needed to count and calculate and that is what they wanted from mathematics. The ancients did not have enough knowledge of the universe to know that the same basic definitions of existence occurred everywhere.

As time went on, discoveries and research became more specialized rather than generalized, moving diametrically in the opposite direction from the primes. The given is the first roadblock to the understanding of the primes. There was so much to learn about the world, the distant lands, the seas, plants, animals, the human body and, the universe to get around to the primes. Even though it is the primes which are the fundamental definitions of what can exist.

There has never been a pressing reason to find the primes. The primes do not hold the cure for disease or enable us to fly. The primes do not unlock the secrets of a horrendous new bomb or a new way to communicate anywhere in the world.

Humans had enough subjects to study to keep them from seeing the core patterns in all the subjects. We tend to notice the differences between various fields of knowledge rather than similarities. The one great similarity is, of course, the manifestation of the same primes.

THE GIVEN

The other great roadblock to understanding the primes is simply the wonders and complexity of the given, our universe. The cosmos can get pretty bizarre at times from an earthling's point of view. There is so much to learn that we have only just begun. Even without a flood of new discoveries, there are all-encompassing things that we have always known of such as gravity and time which still mystify us.

We are finding out more about black holes, vast domains of atoms crushed by intense gravity, where the escape velocity is so high that it exceeds the speed of light and so cannot be seen. Quasars are extremely distant objects that radiate incredible amounts of energy and have most of the people who know much about them either fascinated or baffled or both. In our own solar system, we now know that Neptune has everyday winds that are as powerful as the shock waves from atomic bombs.

Even if no new discoveries were being made about the universe, it's sheer vastness takes a lot of mental energy to attempt to grasp. We can look up at the sky without any kind of scope and see some stars just the way they were when the crucifixion was taking place because the stars are so far away that it has taken that long for light, travelling at the speed of light, to reach us.

Astronomy is only the beginning. Nuclear physics is finding that every atom is a complete and complex solar system. In other branches of science, the twenty-first century is already being called the century of biotechnology. Even with all the great minds in the world, the primes had to wait until someone thought about it and nailed it down or happened across it accidentally due to all the other distractions.

The best know scientific theory of the twentieth century is Einstein's theory of relativity. This theory reveals such things as the slowing down of time and the increase of mass of an object as it nears the speed of light. Time is not absolute, only the speed of light is absolute. On an object moving at the speed of light, the hands of a clock would not move at all and it would seem to an observer in the object, that travelling between any two points in the universe was instantaneous. Even though it may have taken millions of years to observers on earth. There would be no aging when travelling at the speed of light and aging would be drastically slowed down at close to the speed of light.

Travelling at the speed of light is however impossible, at least outside of science fiction. The momentum of an object increases as the mass of the object increases and also as it's speed increases. The mass of any object travelling at the speed of light is infinite. Therefore, to propel it to a higher speed would require an infinite force. Which is of course, impossible. But it does not make the universe that Einstein's theories revealed any less fascinating.

Einstein also believed that gravity was not truly a force like electromagnetism or nuclear binding energy but rather was an innate property of space itself. According to Einstein's theory, space was "warped" by objects with mass much as a bowling ball would warp a mattress and a marble rolling by would fall into "orbit" around the bowling ball.

Also in the twentieth century, Edwin Hubble theorized that the universe is expanding. This implies that the universe began with a great explosion at a certain place and point in time, which we call the big bang, and has been expanding ever since. This theory has been confirmed by the so-called red shift in all galaxies that we observe through a spectroscope.

The red shift is similar to the Doppler effect which occurs when an object that is emitting sound, such as a train, passes an observer standing nearby and the sound drops in frequency as the train passes. This red shift is caused by the light from a distant galaxy "dropping" in frequency to the lower red end of the visible spectrum because the galaxies are all moving relative to each other.

The primes, of course, are beyond the physical universe, existing before the big bang, and so are more primal to reality than either of the above theories. I feel that this has only made the primes more difficult to notice. Maybe if the universe was boring, we might have noticed the primes before now.

The primary goal of physicists now is a "Theory of Everything" to identify a common source for all the known forces. The forces operating the universe are gravity, the electromagnetic force, the strong nuclear force binding atomic nuclei together and, the weak nuclear force that causes radioactivity. Many people, including me, do not believe that gravity is actually a force but rather an innate property of space itself, as described by Einstein's theory.

Some other people believe that there is a mysterious fifth force, which weakly opposes gravity and increases as the density of an atomic nucleus increases. One of the reasons for belief in this fifth force is that iron, which has the densest nucleus although it is not the heaviest element, weighs slightly less than some scientists believe it should.

Whatever you may think of the forces that operate the universe, the Theory of Everything that is being sought is to search for a common source of the forces. The search for the Theory of Everything is just as it's name implies, a search for something that can explain everything. The physicist whose name has been most associated with this search is Stephen Hawking of Oxford. Many people, with whom I agree, believe that the source of everything is simply God.

I also believe that there is an answer to a search for a theory of everything that is a part of the physical realm rather than the spiritual but is at a deeper level than anything that the physicists on the trail of the Theory of Everything are looking for. I believe that I have found it. I have named it the Theory of Primes.

The scientists searching for the Theory of Everything are looking only in the given, the physical universe. Even if they should find an answer, it will still be in the physical universe. The primes are at a deeper level than anything in the physical universe and define what can exist in the physical universe. The opinion of this writer is that the Theory of Primes is the true theory of everything in the physical universe.

PATTERNS AND SPECIALIZATION

The primes combine into many patterns which are manifested in the given. These patterns, such as alphabet, pendulum, pyramid and series, arise from combinations of manifestations of the primes and do an excellent job of camoflaging the true primes. As we have discussed, patterns are dependent on the given for existence while primes are dependent on the given only for manifestation.

Domains are so different from each other that the domain prime, the concept of domainess, does not get much attention. We are so involved with the things that we are dealing with and the differences between them that we do not notice the same primes being manifested. Our given makes possible so many different domains, which have such different applications for human beings that it hides the very concept of the domain prime.

It must be realized that since we are human beings and are intimately involved with the domains that we encounter, we have great difficulty viewing the concept of domain objectively. Cars, mosquitoes, planets, genocidal dictators, tacos, raindrops, shopping malls, songs, police cars looking for speeders, cruise missiles, quarts of Kendall motor oil, the great red spot on Jupiter, gems, psychopaths with guns, cigarettes, members of the opposite sex, newspapers in a language that we cannot understand, pieces of electric wire, Moroccan fez hats, BBC documentaries about exotic creatures on distant islands that few people have ever heard of, inspiring geniuses named Carrie, root systems of willow trees, Persian carpets and, cirrus clouds are all equal manifestations of the domain prime.

The reason that we have such difficulty seeing that is obvious. These domains are so different from our point of view that we never notice what all these domains, and every other domain, has in common, that all are equally manifestations of the domain prime. All domains are identical except for levels, no matter how different the domains may seem to us. Since compensations are composed of changes in domains, this also includes all compensations.

POSTULATE 16: THE ONLY DIFFERENCE BETWEEN ALL DOMAINS, INCLUDING ALL COMPENSATIONS, IS LEVELS.

Missing the primes until now has been the price of specialization. The dividing of science into branches, such as astronomy and biology, leads to specialization and hides the fact that the same primes are manifested in each. We can be sure that there is an infinity of domains, out in the universe as well as in the world all around us, that we fail to notice because we are conditioned to think of the manifestation rather than the prime.

We have been told to remember that all humans regardless of culture and skin color are just as human as we are. As the world has become ever-more international in the twentieth century, this has not always been easy to put into practice. It is likewise not going to be easy to think of all domains as manifestations of the domain prime, rather than as apples, pears and, cars.

An astronomer working at Mount Wilson above Los Angeles will notice that the same physics and forces governs a volleyball game down on the beach as governs the most distant reaches of the universe but does not notice that it manifests exactly the same primes. In our universe, it is generally easier to notice levels rather than domains because there is only a certain amount of the given and this often results in levels tradeoffs when compensations take place. In the primes however, it is most important that domain be understood because it is the first prime and all the other primes revolve around it

WORDS

Speaking of speech, words are possibly the biggest roadblock to the discovery and understanding of primes. If primes are as primal as the Theory of Primes is claiming, we must have been dealing with primes all along without really realizing it. The question words is how we have been handling primes up to know. We should be using primes as a branch of mathematics but instead are dealing with primes very crudely in the domain of language. So crudely in fact, that the primes are not even recognized. The reason for this Theory of Primes is of course, to change this.

Primes are so much a part of the fabric of reality as to be impossible to avoid. The way that human beings have been handling the primes up to now is with the question words such as who, what, where, which, when, why and, how. These everyday words are an inexact and unsatisfactory way of dealing with domain, level, comparison and, compensation. The question words also deal with patterns formed by combinations of primes, which camoflages the basic primes. The words were intended for our everyday convenience, not a way of breaking down reality with mathematical precision.

These words are possibly the major reason that primes have not been discovered until now. The Theory of Primes is that there is only four things that can exist or can happen and that these things should be dealt with primarily as a branch of mathematics, rather than language. The question words have been used for convenience in everyday communication and not for a structured, mathematical analysis of reality as it should be. A method of breaking things down with the precision of the primes is mathematics rather than semantics.

MINDSET FOR DISCOVERY

What is required to understand something like primes is not so much analytical skill as an ability to think outside the box. A large proportion of people who make fundamental discoveries are self-taught as opposed to formally educated. Education imparts knowledge but it also inevitably results in grooved-in thinking to some extent. If we think like everyone else, then we do not notice the things which everyone else does not notice. I have only an Associates degree in mathematics and science and am mostly self-taught. I am certain that if I had a Masters degree or P.H.D., I would not have developed the Theory of Primes for this reason.

It is our personal lives that has a great effect on the operation of our minds and therefore what we are likely to discover that no one has found yet. One of the reasons that I noticed the primes is that I spent many of my formative years working as a busboy, performing such fascinating tasks as sweeping floors. It was frustrating to spend so long walking around picking up things that other people had left behind instead of working a real job.

Now however, it seems that the busboy mentality has been branded onto my mind. I feel as if I can "walk" through the "hallways" of scientific discovery and "pick up" things that others have left behind. The very interesting item that I have just swept up is the primes, the primal branch of mathematics that apparently no one has yet nailed down.

I should also mention that I landed in America in 1968. The lunar missions gave the impression that our world is just a dot when all of space is considered. It also made it appear that the sum of our knowledge might just be a dot when all that there is to know is considered. I wondered if there may be entire new fields of knowledge out there waiting to be found like new planets.

It is also significant that society was not an idyllic paradise in 1968. There was an attitude that if the establishment was right, the world would not be so messed up. Although some of the hippies of the time used terms that were less polite than "messed up". Some people seemed to believe that an idyllic paradise was to be found within, with the assistance of inhaled smoke from the dried leaves of certain tropical plants.

At any rate, I got the message that whatever is accepted as established, unless it is in the Bible which I had already finished questioning, is to be questioned. Although I believed that any questioning should be done using figures of speech that were acceptable to everyone and that the search for improvement should focus on information from the pages of books, rather than smoke from the dried leaves of tropical plants.

I still remember the first book that I ever read all the way through, "Space" by Marian Tellander, a Follett beginning science book for children. It seemed that what America was all about is to find something really grand as far as we could reach and then go for it. I did not get a chance to go to colonize the solar system, making a discovery like the primes will suffice.

CHAPTER SIX

▼

PRIMES IN SCIENCE

Now that we have an understanding of primes, it is appropriate to spend a couple of chapters taking a more detailed look at how complex situations can be broken down into the four basic primes. In this chapter, let's go over the universe from the beginning to where we are today. God set down the primes to guide and define what could exist and what could happen in the universe that he planned to create. When reading this chapter, try to really start seeing the world and the universe in terms of primes.

The seed of the big bang (compensation) was an infinitesimally small (level) point (domain) which physicists (domain) refer to as a singularity (domain). This singularity contained an infinite (level) amount of pure energy (level). The level of the incredible pressure to expand outward (compensation) was much greater (comparison) than the level calling for maintaining the status quo and remaining a singularity (comparison). The singularity (domain) exploded (compensation) violently (level) outward (level)). This explosion (compensation) is known as the big bang and the universe (domain) is still expanding today (level).

In considering primes, we could call the big bang the "primary compensation". Since it is the big bang which is the compensation which began the cycle of compensation that led to every compensation which has ever taken place in our universe (domain).

The first (level) important thing that happened in the big bang (compensation), in the first split second (level), is the separation (domain into sub-domain) of the forces (compensation) from each other. In the first (level) instant (domain), the universe was in such a condition (level) that only one (level) force (domain) was needed or could exist (level). Changing conditions (levels) in the outward (level) explosion (compensation) caused first gravity (domain), then the strong nuclear force (domain) to split off (comparison and compensation creating sub-domains). Then, the weak nuclear force and the electromagnetic force separated (comparison and compensation creating sub-domains).

The next (level) important (level) step (level expressed as domain) in the development (compensation) of the universe (domain) is the mutual annihilation (compensation) of matter (domain) and anti-matter (domain). This was a matter of levels since it was a case of negative charges meeting positive charges and balancing out to zero (level). At this point, it is possible that the universe (domain) was still less than a second old (level).

At maybe three minutes (level expressed in domains) into the universe (domain), we have conditions cool enough (level) for protons (domain) and neutrons (domain) to combine (comparison and compensation) into atomic nuclei (domain). Electrons (domain) are still too energetic (level) to allow complete atoms (domain) to form.

When the universe (domain) was a hundred thousand years old (level expressed in domain), it was cool enough (level) for electrons (domain) with a negative charge (level) drawn by (compensation) the positive charge (level) on protons (domain) to join (comparison and compensation) atomic nuclei (domain) to form atoms (domain). Radiation (domain) separated (comparison and compensation creating sub-domain) from matter (domain).

The standard model for particle physics (domain) that is accepted (domain) today (level) portrays twelve (level) fundamental entities (domains) as the basic (level) constituents (sub-domains) of matter (domain). These twelve (level) entities (domains) are comprised of six (level) quarks (domains) and six leptons (domains).

The name "lepton" (domain) comes from the Greek (domain) word (domain) meaning light and swift (levels). The best-known (level) of these leptons (domain) was the electron (domain). The muon (domain) is a lepton (domain) found in 1937 (domain) as a fundamental (level) component (sub-domain) of cosmic radiation (domain resulting from compensation). The tau (domain) is a lepton that resembles a larger (level) version of the electron (domain) and was discovered in 1977 (domain). The remaining leptons are three types of neutrino (domain), chargeless (level) and apparently massless (level) particles.

Of the twelve (level) particle entities (domain) in the standard model (domain), three (sub-domain) make up matter (domain) as we know it. These are the electron (domain) and two of the quarks (domain). The other particles (domain) exist under only very special circumstances (level), such as in the big bang (domain) or other conditions of extremely high energy (domain expressible in levels). If we consider anti-matter (domain), then the twelve particles (level) would come to twenty-four (level) when the anti-particles (domain expressible in levels) are counted. Some physicists (domain) believe (comparison, compensation) that there exists (domain) gauge particles (domain), which transmit (comparison, compensation) the basic forces (domain) between quarks (domain) and leptons (domain).

The next (level) entity (domain) to appear after matter (domain) is electromagnetic radiation (domain). This radiation is produced (compensation) anywhere (level) in the universe (domain) whenever there is changes (compensation) in atoms (domains), such as electrons (domains) dropping (compensation) to lower energy levels (levels) and giving off energy as a result (compensation). Electromagnetic radiation (domain) is by far (level) the most common (level) compensation (compensation) in the universe (domain).

Electromagnetic waves (domain) vary in frequency (level). Since the frequency (level) of a wave (domain) is inversely proportional (level, comparison, compensation) to it's wavelength (level), we could describe waves by frequency instead of level if that is more convenient (level). At the high-frequency (level) end of the spectrum (domain) is gamma rays and x-rays (level). As we go lower in frequency (level), we find ultraviolet, visible light, infrared, microwaves and, television waves (levels expressed in domain). At the low-frequency end of the spectrum, we find radio waves (level).

The only difference between different types of electromagnetic waves (domain), such as infrared and ultraviolet, is frequency or wavelength. This means that we can express the different kinds of waves as levels, as well as domains. It is similar (comparison) to atoms (domain). The different elements (levels expressed as domains) are expressed as levels since the only difference between different atoms is the number (level) of protons, neutrons and, electrons. However each element, such as gold or iron, can also be expressed as domains.

Electromagnetic waves (domain), as the name implies, are both electric (domain) and magnetic (domain) at the same time (domain). The two fields (sub-domains) making up the wave (domain) are perpendicular (level) to each other. Both fields (sub-domains) contain the energy (level) of the wave (domain) and there are no charged particles (domains) travelling with the field (domain).

Waves (domains) with high frequencies (levels) contain large amounts of energy (level). X-rays (level) for example, are commonly produced (compensation) by firing a beam of electrons (compensation) at a metal target (domain). Light elements (level), such as those found in human flesh and clothing (domain), are poor (level) absorbers of x-rays (comparison, compensation). Heavier elements (level), in contrast (level) stop x-rays (comparison, compensation). This means that x-rays (level) have useful (level) photographic applications (domain) for medicine (domain) and security personnel (domain).

The science (domain) of dealing with visible light (domain expressible in level) is known as optics (domain). It is possible (domain) to divide optics into three branches (sub-domains). Geometrical optics is concerned with behavior of the waves, the wave velocity (level) as well as reflection (compensation) and refraction (compensation).

Physical optics means the properties of light and consists of diffraction (comparison, compensation), interference (comparison, compensation) and polarization (comparison, compensation).

Quantum optics is about the sub-atomic production and effects of the wave and includes atomic excitation (compensation expressible in levels) and the photoelectric effect (comparison, compensation).

The reflection of light (comparison, compensation) can be specular or diffuse (levels). Specular reflection (domain) results from a smooth or shiny surface (sub-domain expressible in levels) such as a mirror (domain) or smooth surface of water (domain).

Generally, shorter wavelengths (level) of light travel slightly slower (level) than longer wavelengths (level). This means that we can break light down into a spectrum or rainbow of color with a glass prism (domain). The slight difference in speed (levels) results in violet (domain expressible in level) having a greater index of refraction (level) than red (domain expressible in level) because violet has the shorter wavelength (level). So, violet deflects (comparison, compensation) at a greater angle (level) than does red. Index of refraction (compensation) of a color (level expressed in domain) of visible light is directly related to (comparison, compensation) it's frequency (level). A diffraction grating (domain) does the same thing as a glass prism (domain) but, it's effect (compensation) with regard to frequency (level) and index of refraction (level) is just the opposite (level).

The prism effect (comparison, compensation) also occurs in nature (domain) when we see a rainbow (domain), which is of course only an optical illusion (level of reality). To see a primary rainbow, the angle between the line between you (domain) and the sun (domain) and the line between you and a cloud of mist (domain) must be around 42 degrees (level). We cannot see a rainbow (domain) when the sun (domain) is more than 42 degrees (level) above the horizon (domain). A complete rainbow is a 360 degree circle (level) but it can only (level) be seen from an aircraft (domain).

Matter (domain) travelling (compensation) through space (domain) is just as much a manifestation (domain) of the compensation prime as electromagnetic radiation (domain) but follows different rules. The difference being that electro-magnetic radiation (domain) always travels (compensation) through empty space (domain) at the speed of light (level) while matter (domain) travels at much lower speeds (level).

In the metric system (domain), a kilogram (level) travelling a meter per second (level) is a unit of momentum (level). A kilogram travelling a meter per second squared is a unit of force called a newton (level). A kilogram travelling a meter squared per second squared gives a newton-meter, also known as a joule (level). A kilogram travelling a meter squared per second cubed gives a joule per second, also known as a watt (level).

When matter (domain) is travelling at very high speeds (level) approaching the speed of light (level), the matter contains a lot of energy (level), manifested as momentum (compensation). As the velocity approaches the speed of light (level), more and more of the energy of momentum (level) is stored as mass (level) rather than velocity (level). This is because (comparison, compensation) the speed of light (level) is an upper limit to velocity (level). It was predicted by Einstein (domain) that an object (domain) travelling (compensation) at the speed of light (level) would have infinite mass (level).

The universe (domain) is an extremely orderly place (level). There are laws (domain) governing all physical processes (comparison, compensation). Hess' law (domain) governs the exchange of energy (levels) during chemical reactions (compensations). Van't Hoff's law (domain) concerns thermodynamics (domain) and states that whenever the temperature (level) of a system in equilibrium (levels) is raised, equilibrium shifts in the direction (compensation) that absorbs heat (level).

Faraday's law (domain) is about the generation of electricity (comparison, compensation) and states that the electromotive force (level) induced by a changing (comparison, compensation) magnetic flux (domain) in a loop of wire (domain) is proportional to the rate of change (comparison, compensation) of magnetic flux (domain) through the coil (domain). The current (comparison, compensation) produced is called induced current. Lenz's law (domain) is about electromagnetism (domain) and states that the direction of an induced current is such that it's own magnetic field (domain) opposes the original change in magnetic flux (domain) that induced the current (comparison, compensation).

After the matter (domain) expanding from the big bang (compensation) had expanded (level) and cooled (level) enough to produce atoms (domain), stars (domain) came into existence as one of the most important domains in the universe (domain). Stars are to be found in a wide variety of forms (levels). The sun could be considered as an average star (domain).

The largest stars are the red giants (sub-domain). A large star, maybe twenty times (level) the size of the sun (domain) turns into a red giant as the hydrogen (domain expressible in level) it has been using for fuel for billions of years (level) is expended. The red giant begins using heavier elements (level), such as silicon (domain expressible in level), as fuel. These heavier elements (level) are fused (compensation) into iron and the star may explode (compensation) into a supernova (domain). Every element in your body (domain) was produced in a star (domain), which ultimately exploded (compensation).

Hydrogen (domain), the lightest element (level), was formed in vast quantities (level) when protons (domain) bonded (compensation) with electrons (domain) about 100,000 years (level) after the big bang (compensation). But, this is the only way for heavier elements (level) such as carbon (domain expressible in level) and oxygen (domain expressible in level) to come into existence (compensation).

In the summer sky (domain), Antares (domain) in the constellation Scorpio (domain) and, in the winter sky (domain), Betelguese (domain) in the constellation Orion (domain) are fine examples of red giants (sub-domain). If either of these stars (domain) were put in the sun's place, planet earth (domain) would be well within the star (level).

Stars (domain) tend to group together (comparison, compensation) in galaxies (domain) due to gravity (domain expressible in levels). There are three (level) fundamental (level) types of galaxy (domain). Our own galaxy and it's larger neighbor, the Andromeda galaxy (domain) are spiral galaxies. Spiral galaxies (domain expressible in level) have a central bulge (sub-domain) and arms (sub-domain) resembling a pinwheel (domain). Elliptical galaxies are somewhat spherical (level) in form with more stars (level) toward the center region (sub-domain). There are also irregular galaxies (domain expressible in level) with a less well-defined shape and which tend to have lower star density (level) than the other two types.

In the early 1920's (domain), at the Lowell observatory (domain) in Flagstaff, Arizona (domain), Vesto Slipher (domain) began to believe (domain) that the galaxies (domains) in the universe (domain) were actually moving apart from each other (compensation). In 1929 (domain), Edwin Hubble (domain) demonstrated (comparison, compensation) that all galaxies (domains) were moving away (compensation) from our galaxy (domain) at a velocity (level) approximately proportional to the distance of the galaxy from us (level). This is due to the expansion (compensation) of the universe (domain) from the primordial big bang (compensation) and somewhat resembles the increasing distance (level) between raisins (domain) in a cake (domain) that is rising (compensation).

The universe (domain) is expanding (compensation) from a primordial explosion (compensation) and every object (domains) in the universe (domain) attracts (compensation) every other object in the universe through gravity. These two forces are opposite (level). This opens the question (comparison) of whether the universe (domain) will ever slow down enough (level) due to gravity (compensation) to fall back (compensation) in toward the location of the big bang (compensation). Much as an object (domain) fired into the sky (domain) ballistically (compensation) will reach a certain height (level) and fall back down (compensation) or, if it is travelling faster (comparison) than the earth's escape velocity (level), will continue (level) indefinitely (level) into space (domain).

The Hubble constant (level) is a figure that states the velocity (level) at which the universe (domain) is expanding (compensation). This brings into play another figure called the deceleration parameter (level), which defines the shape (level) of the universe (domain). The universe (domain) is either open, closed or, flat (levels). The deceleration parameter (level) measures (comparison) how fast (level) the expansion (compensation) of the universe (domain) is slowing down (compensation). An open universe (level) means that the force of gravity (compensation) will never be enough to counter the expansion from the big bang (comparison) and the expansion of the universe will continue forever (compensation). A closed universe (level) is just the opposite, the universe will at some point in the future stop expanding (compensation) and begin contracting (compensation). A flat universe (level) is in between an open and a closed universe and would mean that the expansion is balanced enough with gravity to stop the expansion of the universe (compensation) but not enough to cause the universe to contract (comparison).

The usual way (level) of measuring (comparison) distances (level) in space (domain) from earth (domain) to stars (domains) is by parallax (domain). This is a trigonometric (domain) technique using the orbit (domain) of the earth (domain) around the sun (domain) as the baseline (domain) of a triangle (domain) and measuring (comparison) the slight apparent shift (level) of the target star (domain) against background stars (domain) over a six month (level) period (domain) so that the earth (domain) is on the opposite side (level) in it's orbit (domain). Since the earth (domain) is 93 million miles (level) from the sun (domain), we know that our baseline is twice this distance or, 186 million miles (level). As soon as we can measure (comparison) an apparent shift (level), we can easily (level) calculate (comparison) the distance (level) using the trigonometric functions (domain).

The vast distances (levels) in space (domain) are measured (comparison) in light-years (domain), which is a measure of distance (level), not time (level). Light (domain), or any electromagnetic radiation (domain), travels (compensation) through space (domain) with a velocity of 186, 000 miles per second (level). Or 300,000 kilometers per second (level). A light-year (domain) is the distance (level) thus covered in a year (domain), almost six trillion miles (the U.S. definition of a trillion) (level). A car (domain) travelling (compensation) 55 miles per hour (level) for twenty-four hours a day (domain) would take over twelve million years (level) to drive one light-year (domain).

The trouble (domain) with this technique (domain) is that it is only accurate (level) to about fifty light-years (level) or so. Which in our galaxy (domain) is not much (level), the average (level) star (domain) that you can see (compensation) at night (domain) is possibly seventy light-years distant (level). The stars (sub-domains) of the Big Dipper (domain) are all about this distance (level) from us (domain) and we (domain) could say (compensation) that the Big Dipper (domain) is about 850 million years driving time away at 55 mph (level).

The north star, Polaris (domain), near the Big dipper (domain) is actually a very bright (level) star (domain) very far away (level), about 650 light-years (level). If you were near (level) Polaris (domain), this means that you would see (compensation) the Big Dipper (domain) in the opposite direction (level) in the sky (domain) relative to earth (domain) and it would be much smaller (level) than we see it from earth (domain) due to greater distance (level). Somewhere below (level) the bowl (sub-domain) of the Big Dipper (domain) viewed from Polaris (domain), you (domain) would see an average yellowish (level) star (domain) known as the sun (domain).

The north star (domain) happens to be a cepheid (sub-domain). These stars are significant (level) to people (domain) on earth (domain) who want (comparison) to measure (comparison) the distances (level) in space (domain) because (comparison) the stars (domain) vary (compensation) in brightness (level). Cepheids (sub-domain) that are larger (level) and brighter (level) have a longer cycle (level) of variation (compensation) in brightness (level). The relationship (comparison, level) between brightness (level) of a cepheid and the length (level) of it's brightness (level) cycle (comparison, compensation) is closely proportional (level, comparison). This makes cepheids very useful (level) indicators (comparison) of distance (level) far beyond (level) that made possible (domain) by parallax (domain). It would be much the same (level) if all light bulbs (domain) were of the same brightness (level) and you (domain) could judge (comparison) the distance (level) of a light (domain) by it's apparent brightness (level).

A similar (level) technique (domain) for approximating (comparison) distances (level) to far away (level) galaxies (domains) is known as the Tully-Fisher relation (domain). The principle (domain) of this measurement technique (domain) is simply that more luminous (level) galaxies (domain) tend to be more massive (level) than dimmer (level) galaxies (domains) and so rotate (compensation) slower (level). A comparison is made between the apparent brightness (level) of a galaxy (domain) and the rotation (compensation) rate (level).

Since galaxies (domains) rotate (compensation) much too slowly (level) to be observed (compensation) over a human lifetime (domain), measurement (comparison) of rotation (compensation) speed (level) is done (compensation) by measuring redshift (level) of the part (sub-domain) of the galaxy (domain) that is moving away (compensation) from us (domain) and blueshift (level) of the part (sub-domain) of the galaxy (domain) that is moving (compensation) toward (level) us (domain). Much the same (level) as looking (compensation) at a helicopter (domain) on the ground (domain), one blade (domain) is moving (compensation) toward (level) you (domain) while (domain) the other is moving away (level).

As we (domain) look (compensation) further out (level) into the universe (domain), things only get more interesting (level). In 1973 (domain), astronomers (domain) discovered (domain) gamma ray bursts (compensation expressible in domain). It is now (level) believed (domain) that these bursts (domain) are along the same lines (level) as solar flares (compensation expressible in domain) but come from (compensation) neutron stars (domain).

These neutron stars (domain) are essentially large (level) spheres (level) of crushed (level) atoms (domains) where the negatively-charged (level) electrons (domain) and positively-charged (domain) protons (domains) have repelled (compensation) each other off and left only a large (level) body (domain) of neutrons (domain) held together (compensation) by gravity (level express-ible in domain).

Since an atom (domain) is normally (level) empty (level) space (domain), the mass (level) of neutron stars (domain) are extremely high (level). Being accustomed (domain) to earth (domain), we cannot (domain) imagine the density (level) of a neutron star (domain). The gamma ray bursts (domain) are thought to be (domain) caused (compensation) by electrons (domain) caught in (compensation) intense (level) magnetic fields (domain).

On our more (level) familiar (domain) earth (domain), we find (compensation) a very special (level) chemical known as water (domain). There is a complex (level) interplay (comparison, com-pensation) involving water (domain) between land (domain) and ocean (domain). Water vapor (domain) evaporated (compensa-tion) from oceans (domain) falls (compensation) on land (domain) as precipitation (compensation) while the oceans (domain) are replenished (level, compensation) by rivers (domain) which have drained (compensation) watersheds (domain) on land (domain). The earth's land (domain) gets about 26% of the total (level) precipitation (compensation) but contributes only 16% (level) of the total evaporation (compensation). The earth's oceans (domain) receive 74% (level) of the precipitation (compensation) while contributing 84% (level) of the evaporated water (domain).

The earth's (domain) oceans (sub-domain) accumulate (compensation) salt (domain) due to water (domain) wearing down (compensation) and dissolving (compensation) rocks (domain) and minerals (domain). The Red Sea (domain) is the saltiest sea (level) in the world (domain) due to intense (level) evaporation (compensation). Sometimes, masses of water (domain) are of different density (level) due to different salinity (level) or, concentration (level) of salt (domain) and this causes movement (compensation) of the water (domain). Water (domain) in the ocean (domain) is also moved (compensation) by winds (compensation) on the surface (domain).

There are ocean currents (compensation), such as the Gulf Stream (domain) due to the Coriolis effect (domain). This effect is caused by the rotation (compensation) of the earth (domain), in the northern hemisphere (domain), ocean currents (compensation) are deflected clockwise (level), in the southern hemisphere, counterclockwise (level).

This precipitation (domain, compensation) which falls (compensation) on land (domain) is vital (domain) to human beings (domain) for fresh (level) drinking water (domain) and for farming irrigation (domain).

There is a wide variety (level) of farming soils (domain) on the earth (domain).

Sandy (level) soils (domain) offer the advantage (level) of being well drained (domain expressible in level), accessible to air (domain) and, allow easier root (domain) penetration (compensation). However the ease (comparison) with which water (domain) can penetrate (compensation) sandy soils (domain) causes organic matter (domain) and nutrients (domain) to get leached out (compensation). This means that more (level) fertilizer (domain) is often necessary (domain) in sandy soils (domain expressible in levels). Sand (domain) is produced (compensation) by water (domain) and so, sandy soils are found (domain) in land (domain) that was part (domain) of a body of water (domain) millions (level) of years (domain) ago.

Clay soils (domain expressible in levels) are just the opposite (level) of sandy soils (domain expressible in levels). Nutrients (domain) are usually plentiful (level) but, plant roots (domain) may have difficulty (level) penetrating (compensation) and clay soils (domain) are not as easy (level) to plow (compensation) as sandy soils (domain). The worst (level) disadvantage for agriculture (domain) of clay soils (domain) is the slow (level) water drainage (compensation). Silty soils (domain expressible in levels) are not good (level) for farming (domain) due to poor (level) drainage (compensation), lack of nutrients (level) and, susceptibility (level) to erosion (compensation).

The type of soil (domain) known as loam (domain) is a farmer's dream (domain) because it provides a very workable (level) combination (domain) of all of the above soil types (domain). It has the best (level) advantages of all three (level) soil varieties (domain).

The earth (domain) is heated (compensation) unevenly (level) by the sun (domain) due to the fact that water (domain) and land (domain), and different types of land (sub-domain), absorb and hold (compensation) heat (compensation) at different rates (level). The actions (compensation) of the atmosphere (domain) with regard to land (domain) at any given point (level) in time (domain) is known as weather (domain). These actions of the atmosphere in a given geographic location (level) over the course of a year (domain) is known as climate (level).

The shape (level) of the earth (domain) is spherical (level) and so, the earth (domain) is warmed (compensation) less at the poles (level) than at the equator (level), due to the angles (level) of the sun's incoming (compensation) rays (domain). Warm (level) air (domain) tends to rise (compensation) while cold (level) air (domain) tends to sink (compensation). The sun (domain) does not (comparison) warm (compensation) the air (domain) directly but warms (compensation) the earth (domain) by radiant (compensation) energy (level). As warm (level) air (domain) at the equator (level) rises (compensation) holding (domain) water (domain) it flows (compensation) toward the pole (level), cooling (compensation) and drying (compensation) as it goes (domain). When the air mass (domain) reaches about 30 degrees (level), or a third (level) of the way to the pole(level), it sinks (compensation) and warms up (compensation) and flows back (compensation) toward the equator (level).

Such flowing (compensation) of air (domain) on both sides (domain) of the equator (level) are known as Hadley cells (domain). This pattern (domain) of air (domain) flow (compensation) causes the dry (level) lands (domain) to be found all over (level) the world (domain) at around 30 degrees north (level) and south (level) of the equator (level). The air (domain) is falling (compensation) at these latitudes (level) and prevents (compensation) local air (domain) from rising (compensation) to form (compensation) clouds (domain).

Ocean currents (compensation) can (level) also have a profound (level) effect on climate (level). Britain (domain) is at the same latitude (level) as icy (level) Labrador (domain) in Canada (domain). Fortunately (domain), the Gulf Stream (compensation) brings warm (level) water (domain) to the area (domain) from the tropics (domain). There is even an area in Cornwall (domain) where palm trees (domain) can grow (compensation). The Gulf Stream (compensation) has a similar (level) but, slightly lesser (level) effect on southern (level) Norway (domain) and, Sweden (domain). There are deserts (domain, level) along the coast (domain) of Chile (domain) and Namibia (domain) caused (compensation) by cold (level) ocean (domain) currents (compensation) preventing (compensation) water (domain) from evaporating (compensation) to form (compensation) clouds (domain).

Thunderstorms (domain) result (compensation) from powerful (level) air (domain) currents (compensation) inside (domain) towering (level) cumulus clouds (domain). Warm, wet (level) air (domain) rises (compensation) until it reaches a cool enough height (level) to condense (compensation) into water droplets (domain). The water droplets (domain) may then freeze (compensation) into ice crystals (domain). Each change (compensation) of state of matter (level) releases (compensation) energy (level) into the cloud (domain). When the water droplets (domain) or ice crystals (domain) grow large (level), they begin to fall (compensation). While falling, the droplets brush against uprising (compensation) air (domain) producing (compensation) static (level) electric charges (domain) much like (comparison) walking (compensation) across a carpet (domain). When the charge (domain) builds up enough (level), it is discharged as lightning (compensation).

Snow (domain) is formed (compensation) by sublimation (compensation) of water vapor (domain), the changing (compensation) of a vapor (level) to a solid (level) by a drop (compensation) in temperature (level). Snow (domain) begins in the ice crystals (domain) making up high (level) cirrus clouds (domain). Rising (compensation) water vapor (domain) causes the crystals to grow (compensation). If the crystals (domain) pass through (compensation) stratus clouds (domain), there is even more (level) growth.

Without radiation (domain) from the sun (domain), earth (domain) would be dark (level) and frozen (level). The temperature (level) would be close to absolute zero (level). In this condition (domain), all atomic and molecular motion stops (compensation). Since heat (compensation) is movement (compensation) of atoms (domain) and molecules (domain) which slows down (compensation) as the object (domain) gets colder (compensation), there must be a point (level) at which all motion stops (compensation). We call this point absolute zero (level). We reach this point at-459 degrees Fahrenheit or-273 degrees Celsius (level). Absolute zero is also known as zero degrees Kelvin. A Kelvin degree (domain) is the same size as a Celsius degree (domain) so that zero degrees Celsius would be 273 degrees Kelvin (level). Even though the interiors of stars (domain) are as hot as millions of degrees (level), the universe (domain) on the whole is a very cold (level) place (domain) with an average temperature of 3-4 degrees Kelvin (level). Nowhere on earth (domain) comes remotely close to this (level).

Interestingly, when we reach absolute zero (level), time (domain) ceases to exist. As we have seen, for all practical purposes, time is motion (compensation). If there is no motion, then there is no meaningful definition of time. This has no relationship to Einstein's theories concerning the relativity of time (domain).

The reason that even the coldest (level) regions (domain) on earth (domain) comes nowhere near (level) absolute zero is radiation (domain) from the sun (domain). This radiation (domain) originates deep (level) in the sun (domain) as a result of fusion (domain), the same process as in a hydrogen bomb (domain). The extreme heat energy (level) in the sun (domain) fuses (compensation) four (level) atoms of hydrogen (domain) into one atom of helium (domain). This leaves just a little bit (level) of mass left over, which is converted (compensation) into energy (level). Millions of tons (level) of helium (domain) is produced every second (domain) in the sun (domain) by this process (domain), which frees (compensation) an incredible amount of energy (level). The sun's radiant energy (domain), travelling (compensation) at the speed of light (level), takes about eight minutes (level) to reach the earth (domain).

In the earth's atmosphere (domain), this radiation (domain) is absorbed (compensation) by water vapor, ozone, carbon dioxide, particles of dust and, ice crystals (all domains). Much (level) of the sun's energy (level) is reflected (compensation) back into

space (domain) by cloud cover (domain) and by the earth's surface (domain). Oceans (domain) and black asphalt (domain) in
parking lots (domain) absorb (compensation) almost all (level)
the radiation (domain) received while white snow (domain)
reflects (compensation) most (level) of it. The thicker (level) the
cloud type (domain), the more radiation (level) is reflected back
(compensation) into space (domain). Cirrus (domain) reflects
back (compensation) the least (level), cumulus (domain) the
most (level). The amount (level) of incoming (level) radiation
(domain) that a planet (domain) or other body (domain) reflects
back (compensation) into space (domain) is known as the albedo
(level) and is usually expressed as a percentage (level). A dull
black (level) object (domain) will have an albedo near zero (level)
while a very shiny (level) object (domain) has an albedo close to
100.

The sky (domain) on earth (domain) appears as light blue
(level) because of the scattering (compensation) of sunlight
(domain). Some colors (domains expressible as levels) are scattered (compensation) by oxygen (domain) and nitrogen atoms
(domain) and molecules (domain) more than (level) others and
when added together we get the familiar (level) sky blue.

The sky (domain) is actually closer (level) to violet but our
eyes are not very sensitive to violet so we see blue. Aside from the
gas atoms (domains), larger particles (domain) scatter (compensation) radiation (domain) of wavelengths (level) toward the red
end (level) of the spectrum (domain) and with enough (level)
dust (domain), the intensity (level) of the blue will be dimin-
ished (compensation). The sky (domain) appears red at sunset
because the rays (domain) have to travel (compensation)
through a larger volume (domain) of sky due to the angle (level)
and so are scattered (compensation) more than (level) when we
see blue. I (domain) have been in (domain) cities (domain)

where parts of the sky (domain) appear a yellowish (level) brown due to the different (level) refraction (compensation) of sunlight (domain) by oxides of nitrogen (domain) produced by (compensation) industrial processes (domain).

A promising way of using (compensation) the energy (level) of the sun (domain) to produce (compensation) electricity (compensation) is the solar pond (domain). Israel (domain), lacking oil (domain) but having plenty (level) of sunlight (domain), is especially interested in solar ponds (domain). The pond (domain) is a deep (level) basin (domain) and the water (domain) at the bottom (level) of the pond (domain) is given a greater salinity (level) than the water (domain) at the top (level). The bottom (level) of the pond (domain) is dark (level) and the sun's rays (domain) striking (compensation) the floor (domain) of the pond (domain) heat up (compensation) the water (domain) at the bottom (level). Since salt water (domain) is heavier (level) than fresh water (domain), the bottom water (domain) cannot rise up (compensation) and mix (compensation) with the cooler (level) water at the top (level) as would happen (compensation) in a swimming pool (domain) with a dark (level) floor. The water (domain) at the bottom (level) of the solar pond (domain) becomes hotter and hotter (level) until it is piped away (compensation) and used for energy (level) and the process (domain) is started over.

The radiation (domain) from the sun (domain) is not all that reaches (level) earth (domain) from space (domain). Cosmic rays (domain) bombard (compensation) the earth (domain) constantly (level). Cosmic rays (domain) consist mostly of protons, neutrons, electrons and, gamma rays (all domains) and come from various sources (domain), including the sun (domain). Cosmic rays (domain) may contain a vast amount (level) of energy (level). Many (level) of the particles (domain) that reach

the earth's surface (domain) are secondary (level), rather than
primary (level), cosmic rays (domain). The primary rays collide
with atoms (domain) in earth's atmosphere (domain) and give
off (compensation) secondary particles. Many (level) of these
particles (domain) are muons (domain), heavier (level) relatives
(domain) of electrons (domain).

When particles (domain) are travelling (compensation) close
to (level) the speed of light (level), relativity comes into play
(domain). Many (level) particles (domain) are unstable (level)
and decay (compensation) in a certain amount (level) of time
(domain). But, that is only if the particle (domain) is stationary
(level). When travelling (compensation) at relativistic speeds
(level), time (domain) drastically slows down (compensation)
and particles (domain) decay (compensation) by this clock
(domain).

One important (level) effect (compensation) of cosmic rays
(domain) is production (compensation) of carbon-14 (domain
expressed in level), which is so important (level) to archeologists
(domain). Carbon (domain) usually has six protons (level) and
six neutrons (level) in it's nucleus (domain). Nitrogen (domain),
which makes up 78% (level) of earth's lower (level) atmosphere,
or troposphere (domain), has seven protons and seven neutrons
(levels) in it's nucleus. When a cosmic ray neutron (domain)
strikes (compensation) a nitrogen nucleus (domain), it absorbs
(compensation) the neutron (domain) but knocks out (compensation) a proton (domain).

The resulting (compensation) atom is carbon-14 (domain),
meaning that it has two extra (level) neutrons (domain) compared with ordinary carbon-12 (domain expressed in level).
Carbon-14 is so useful because it is not completely stable (level)
and eventually decays (compensation) to ordinary carbon-12
(domain expressed in level).

Since all organic material (domain) contains carbon (domain) and since the proportion (level) of carbon-12 to carbon-14 is virtually always constant (level), all that is necessary to determine the age (level) of an archeological artifact (domain) is to measure the ratio (level) of carbon-14 to carbon-12 in a sample of the artifact (domain). Carbon-14 has a half-life of 5600 years (level), meaning that of a given number of carbon-14 atoms, half will decay to carbon-12 in 5600 years (domain). This gives it an ideal range (domain) for dealing with artifacts (domain) from human history (domain). Carbon-14 dating (domain) is accurate to approximately fifty thousand years (level).

Some (level) of the visitors (domain) that the earth (domain) gets from space (domain) can be downright destructive (level). Sixty-five million years ago (level expressed in domains) all or part or a comet (domain), with it's rocky nucleus (domain) hit (compensation) the earth (domain) at tremendous speed (level). The force (level) of the explosion (compensation) made all the nuclear bombs (domain) in the world (domain) today (level) look like (comparison) firecrackers (domain). The shock wave (compensation) that went around the world (domain) from the blast (compensation) contained (domain) so much energy (level) that everything (domain) it passed was incinerated (compensation). Much of the earth's surface (domain) and atmosphere (domain) was thrown (compensation) into space (domain) as debris (domain). Tremendous (level) earthquake (compensation) and volcanic activity (compensation) transformed (compensation) much of (level) the earth's surface (domain). The waters (domain) of the earth's oceans (domain), much of it boiling (compensation) was splashed down (compensation) all over the earth (domain).

CHAPTER SEVEN

▼

PRIMES IN TECHNOLOGY AND HUMAN AFFAIRS

Technology is the use of scientific knowledge to modify and improve our environment. In this chapter, we will become more familiar with the primes manifested all around us. We will see once again that exactly the same four primes are manifested in technology and human affairs as are manifested in nature.

While we are busy (compensation) finding out (domain) more (level) about the universe (domain), we (domain) are hoping to be able to travel (compensation) deeper (level) into space (domain) someday (domain). Space (domain) travel (compensation) to other planets (domain) in our solar system (domain) resembles (comparison) electrons (domain) in different orbits (domain) in an atom (domain). Venus (domain) is closer (level) to the sun (domain) than earth (domain) and is therefore at a lower energy level (level), just as electrons (domain) in atoms (domain) are at a lower energy level (level) when closer (level) to the atom (domain). If we launch (compensation) a spacecraft (domain) in the direction opposite (level) to that which the earth (domain) is moving (compensation) around the sun (domain), the spacecraft (domain) will still be going in the same direction (level) as the earth (domain) except slower (level). When the sun (domain) is rising (compensation), the direction (level) the earth (domain) is travelling (compensation) through space (domain) is straight above (level) your head (domain).

The spacecraft (domain) will have less momentum (level) than earth (domain) but will still be in the same orbit (domain expressible in level) as earth (domain). The earth (domain) has just the right amount (level) of momentum (level) to remain (compensation) in it's orbit. The spacecraft (domain), having less (level) than this amount of momentum (level), will be pulled in (compensation) by the sun's gravity (domain) while continuing in the same direction (level) as earth (domain). If everything (domain) is timed just right (level), the spacecraft (domain) will reach the orbit of Venus (domain expressible in level) just as the planet (domain) reaches that point (level) in it's orbit. In all the time (level) that it takes the spacecraft (domain) to reach Venus (domain), it may spend only ten minutes (level) with it's engines (domain) on. Momentum (level) handles the rest.

One of the most difficult (level) aspects of manned space flight (domain) is reentry into earth's atmosphere (domain). The atmosphere (domain) must be entered at just the right angle (level). There is an angular window (domain) of only about 2 1/2 degrees (level) that the spacecraft (domain) must hit. If the craft (domain) enters earth's atmosphere (domain) at too steep an angle (level), it will certainly be incinerated (compensation) by frictional heat (compensation). If it enters at too shallow an angle (level), it will bounce off (compensation) the atmosphere (domain) like a flat stone (domain) skipping (compensation) over water (domain). Even when reentry (domain) is done properly (domain), a tremendous amount (level) of heat (compensation) is generated (compensation) that must be gotten rid of (compensation). This can be done by layers (levels) or tiles (domains), which get extremely hot (level) and then peel off (compensation), exposing another layer (domain). Cooling in this way is known as ablative cooling (compensation). On the Apollo spacecraft (domain), the blunt bottom (level) of the craft (domain) created a shock wave (compensation) in the atmosphere (domain) which expended much of the energy generated as heat (level).

Nuclear bombs (domain) are one branch (sub-domain) of technology (domain) that has gotten a lot of attention (level) in the twentieth century (domain). All atomic nuclei (domain) use the strong nuclear force (domain) to hold together (compensation). This is necessary because positively-charged (level) protons (domain) naturally repel each other (compensation) and require a binding force (domain) that can overcome (level) this repulsion (compensation). Even with neutrally-charged (level) neutrons (domain) in the nucleus (domain) diluting (level) the mutual repulsion (compensation) of the positive (level) charges (domain).

Atomic bombs (domain) operate (compensation) on the principle (domain) that there are nuclei (domain) of certain heavier (level) elements (domain) that can be split (compensation) by a neutron (domain) travelling (compensation) at high velocity (level) in which further neutrons (domain) will be released by the splitting (compensation) which will split (compensation) other nuclei (domain). The atoms (domain) of these heavier elements (level) split (compensation) into two lighter (level) atoms (domain) during this process (domain). The significant (level) factor (domain) is that the two (level) nuclei (domain) of the resultant lighter (level) atoms (domain) contain (domain) a little less (level) binding energy than the original large (level) atom (domain). This energy (level) has to go somewhere (compensation), since energy (level) cannot be created (compensation) or destroyed (compensation), but only changed in form (compensation). The energy (level) is released as heat (compensation) and is where the tremendous power (level) of the bomb (domain) comes from. Energy (level) and mass (domain) are interchangeable (compensation). According to Einstein (domain), a little bit of mass (level) is equal to (comparison) a tremendous amount of energy (level). When the large (level) atom (domain) is split (compensation) into the two smaller components (sub-domain), we find that a little bit of mass (level) has disappeared (domain).

Simple burning and conventional explosives (compensation) operate on a similar principle. The heat from burning (compensation) comes from the breaking of molecules (domain) and the resulting release (compensation) of the energy (level) of the bond between the molecules (domain). This is why most fuels (domain), such as gasoline (domain) or kerosene (domain), consist of complex (level) molecules (domain). Fuels usually contain carbon (domain) because of it's amazing ability to build complex (level) molecular structures (domain). This is the same reason that life (domain) is carbon-based (domain). Gasoline (domain) is a very complex molecule and it is transformed (compensation) into carbon monoxide (domain), which is a very simple (level) molecule. Energy that was binding (compensation) the component molecules (sub-domains) of gasoline (domain) is released as heat (compensation) so that you can drive to work (compensation).

Unlike fuels (domain), conventional explosives (domain) are usually based on nitrogen (domain), rather than carbon compounds (domain). Nitrogen (domain) is very unreactive (comparison) and it takes a lot of energy (level) to get it to form compounds (domain) with other elements (domain). Once it does combine (compensation), the bonds (domain) therefore contain a lot of energy (level). When these bonds are broken by heat (compensation), energy is released very rapidly and we get a bang (compensation). However, a nuclear explosion (compensation) uses the energy binding (compensation) the nucleus (domain) itself, not just the molecular bonds (domain) between atoms (domain).

In a nuclear reaction (domain), we need to have what we call a chain reaction (compensation cycle expressed in domain). When a neutron (domain) splits (compensation) an atom (domain), we need at least one neutron (domain) to be released in order to split (compensation) other atoms (domain). In actual practice, one neutron (domain) is not enough (level) to sustain the chain reaction (domain) since some neutrons (domain) will escape (domain). During a sustainable chain reaction, 2.5 (level) neutrons (domain) are released (compensation) for every neutron (domain) that is split (compensation). The original neutron (domain) must obviously have sufficient velocity (level) to split the atom (domain).

Only two (level) materials (domain) are known (domain) which will suffice (level) for a bomb (domain), plutonium (domain) and an isotope (level) of uranium U-235 (domain expressed in level). A certain amount (level) of the material (domain) must be present (domain) in the right conditions (level) for the chain reaction (compensation expressed in domain) to occur. This minimum amount necessary (level) is known as the critical mass (level). The critical mass (level) is necessary because too many (level) of the high-speed (level) neutrons (domain) will escape (domain) from the material (domain) before splitting (compensation) another nucleus (domain) if the material (domain) is below a certain size (level).

The most important (level) choice (comparison) to be made (compensation) when building (compensation) the bomb (domain) is whether to use plutonium (domain) or U-235 (domain). Plutonium is much more easily available (level) but is more difficult (level) to make (compensation) a bomb (domain) with. The bomb (domain) which exploded (compensation) on Hiroshima (domain) in 1945 (domain) was a uranium bomb (domain), the one on Nagasaki (domain) was a plutonium bomb (domain). U-235 is so-called because most natural (domain) uranium (domain) has 238 (level) nucleons (domain) in it's nucleus (domain), 92 protons and 146 neutrons (levels of domains). However, this will not (domain) split (compensation) under the impact (compensation) of a high-speed (level) neutron (domain) because the three (level) extra neutrons (domain) hold it together (compensation) more tightly (level). But, an isotope (level) of uranium (domain), U-235 (level of domain), with three (level) less (domain) neutrons (domain) in the nucleus (domain) will split (compensation).

In a bomb (domain), detonation (compensation) is achieved by making a sub-critical (level) mass (domain) critical (level), usually with conventional (level) explosives (domain) shooting (compensation) two (level) smaller (level) masses of U-235 (domain expressed in level) together (domain) in the uranium bomb (domain). In the more difficult (level) plutonium bomb (domain), the plutonium (domain) would have to be fashioned (compensation) into a virtually perfect (level) sphere (domain expressed in level) maybe the size (level) of a tennis ball (domain). About 800 pounds (level) of perfectly packed (level) explosives (domain) around (domain) the plutonium (domain) would explode (compensation) inward (level), compressing (compensation) the sub-critical (level) plutonium (domain) sphere (level) into a critical mass (level).

The plutonium (domain) would have a layer (domain) of U-238 (domain expressed in level) around it (domain) to reflect (compensation) escaping (domain) neutrons (domain). The surrounding (domain) conventional (level) explosives (domain) have to be detonated (compensation) with perfect timing (level), using electrical capacitors (domain) for coordination (level). As you can see, the uranium bomb (domain) is much simpler (level). Complicating (compensation) the plutonium bomb (domain) still further (level) is the fact (domain) that plutonium (domain) is the most toxic (level) substance (domain) known to man (domain). In the chain reaction (domain), the nucleus (domain) of uranium (domain) or plutonium (domain) is split (compensation) into atoms (domain) like barium (domain) and krypton (domain). In either (domain) bomb (domain), the chain reaction (domain) is initiated (compensation) by a small (level) source (domain) of neutrons (domain) inside (domain) the bomb (domain).

The biggest (level) concern in building (compensation) a nuclear bomb (domain) is obtaining (domain) whichever material (domain) you have chosen (comparison), U-235 (domain) or plutonuim (domain). Neither occurs readily (level) in nature (domain). Which is why everyone (domain) does not have (domain) atomic bombs (domain).

To build (compensation) a uraniun bomb (domain), one uranium atom (domain) in 140 on average (level) in a lump of uranium-238 (domain) is the U-235 isotope (domain expressed in level) that is needed (domain). The only way to get it (domain) is to separate it out (compensation). Scientists (domain) have figured out (compensation) three (level) ways (domain) to separate U-235 from U-238 (domains). In the electromagnetic method (domain), the uranium (domain) is shot (compensation) between (level) the poles (domain) of a magnet (domain) after being vaporized (compensation) and the heavier (level) U-238 (domain expressed in level) is deflected (compensation) more (level) and so, separated (compensation) from the U-235 (domain). In the centrifuge method (domain), the vaporized (level) uranium (domain), combined with (domain) fluorine (domain) as uranuim hexafluoride (domain), is whirled around (compensation) and the heavier (level) atoms (domain) swing to the outside (compensation) more (level). In the diffusion method (domain), more light (level) U-235 (domain) seeps (compensation) through a porous divider (domain) than heavier (level) U-238 (domain) and so, separation (compensation) is possible (domain).

Plutonium (domain) is an entirely man-made (domain) element (domain) that is not known (domain) to occur in nature (domain) at all. It has (domain) 94 protons and 145 neutrons (levels of domains) in it's nucleus (domain), making it slightly heavier (level) than uranium (domain). Plutonium (domain) is actually made (compensation) slowly (level) from ordinary U-238 (domain). A neutron (domain) moving (compensation) at a slow enough speed (level) can be captured (domain) by the nucleus (domain) of uranium (domain). When U-238 (domain) absorbs (compensation) a neutron (domain), it becomes unstable (level), radioactive (level) U-239 (domain). This isotope (level) of uranium (domain) is also unstable (level) and releases (compensation) a beta particle (domain) from the nucleus (domain), changing (compensation) it to the element neptunium (domain), with 93 protons (level of domain). Neptunium (domain) lacks (domain) stability (level) and gives off a beta particle (domain) to become (compensation) the plutonuim (domain) that is sought (domain).

There is three (level) types of radioactivity (compensation expressed in domain). Alpha, beta and, gamma (domains). Alpha (domain) is positively charged (level) protons (domain). Gamma (domain) is extremely high frequency (level) and high energy (level) electromagnetic waves (domain). A beta particle (domain) is a high-energy (level) electron (domain) given off (compensation) when a neutron (domain) changes (compensation) into a proton (domain), which happens (compensation) only under very special conditions (level expressible in domain).

This nuclear bomb (domain) that we have been discussing has been merely an ordinary (level) atomic bomb (domain). Devastating (level) enough as it is, but falling far short (level) of the power (level) of a hydrogen bomb (domain), which is what virtually all (level) of today's (domain) nuclear weapons (domain) are. An atomic bomb (domain) uses the process of fission (domain), or splitting (compensation) of a nucleus (domain), to release (compensation) the binding energy (domain) which was holding (compensation) the nucleus (domain) together (domain) against the mutually repulsive (level) forces (domain) of the positively charged (level) protons (domain). There is another (domain) process, known as fusion (domain), which releases (compensation) still more energy (level). Fusion (domain) gives off (compensation) energy (level) because the energy necessary (level) to bind (compensation) four (level) heavy (level) hydrogen atoms (domain) together (domain) is less (level) than that required (domain) to bind (compensation) one (level) helium atom together (domain). As we saw earlier (domain), this is the process (domain) used to generate (compensation) energy (level) in the sun (domain), except that ordinary (level) hydrogen (domain) instead of heavy (level) hydrogen is used in the sun (domain). Hydrogen (domain) is the lightest (level) element (domain) with a single proton (domain) in the nucleus (domain) while, heavy (level) hydrogen also has a neutron (domain) in the nucleus (domain).

Heavy (level) hydrogen is known as deuterium (domain) and is rare (level) and expensive (level). In the hydrogen bomb (domain) it is used in the form of heavy water (domain expressed in level), which is water (domain) with it's molecules (domain) composed of (domain) atoms of hydrogen (domain) with a neutron (domain). As opposed to ordinary water (domain), in which the hydrogen atoms (domain) contain no neutrons (domain). In this fearsome (level) bomb (domain), a layer (domain) of heavy water (domain) surrounds (domain expressed in level) an ordinary (level) atomic bomb (domain), which acts as a mere (level) detonator (domain). At a temperature (level) of millions of degrees (level), which the explosion (compensation) of the atomic bomb (domain) provides, four (level) atoms of heavy hydrogen (domain) fuse (compensation) into one (level) of helium (domain) and release (compensation) the extra binding energy (domain) as heat (compensation expressed in level) in the process (domain). The result is devastation (compensation).

When destroying (compensation) a city (domain), the bomb (domain) can be detonated (compensation) either at ground level (domain) or at a certain altitude (domain expressed in level). An altitude (domain) detonation will spread destruction (compensation) over a wider (level) area (domain) as the fireball (domain) scorches (compensation) the ground (domain). The ground detonation (domain, compensation) however, will throw (compensation) more radioactive (domain) debris (domain) and dust (domain) into the sky (domain), which will later (domain) spread (compensation) radiation (domain) over a wide (level) area (domain). Radioactive (level) dust (domain) will form (compensation) condensation nuclei (domain) for raindrops (domain) which will spread (compensation) radiation (domain) throughout (level) the watershed (domain).

One effect (compensation) of a detonation (compensation) high (level) in the atmosphere (domain) would be the effect known as electromagnetic pulse (domain). A high-altitude (level) nuclear blast (domain) would generate (compensation) very energetic (level) gamma radiation (domain) which would strip (compensation) large numbers (level) of electrons (domain) from atoms (domain) of nitrogen (domain) and oxygen (domain) in the atmosphere (domain). Guided by (compensation) the earth's magnetic field (domain), the liberated (domain) electrons (domain) move (compensation) in large (level) spiral (level) paths (domain) due to the cyclotron effect (domain), and in so doing, emit (compensation) a short (level) but, extremely powerful (level) pulse of electromagnetic waves (domain). The frequency (level) of this pulse (domain) extends from 10 khz to 100 mhz (domain expressed in levels), exactly the same (domain) range of frequency (domain) used by the majority (level) of the world's communication systems (domain). This pulse (domain) would destroy (compensation) communications equipment (domain) by inducing (compensation) destructive electric fields (domain) and arcing (compensation) within the equipment (domain). The very factors (domain) making digital (level) equipment (domain) more powerful (level) and useful (level) make it ever-more sensitive (level) to destruction (compensation) from electromagnetic pulse (compensation expressed in domain).

There is, of course, peaceful (domain) uses (domain) for nuclear energy (domain). A nuclear reactor (domain) is used to generate (compensation) power as well as producing (compensation) radionuclides (domain). In a nuclear reactor (domain), a controlled (level) reaction (domain) takes place (compensation), unlike (level) in the bomb (domain). For a reactor (domain), we want to slow down (compensation) the fast (level) neutrons (domain) produced (compensation) by fission (domain) using a moderator (domain), which must not absorb (compensation) the neutrons (domain) but merely reduce (compensation) the kinetic energy (level) of the particles (domain). One of the best moderators (domain) is carbon (domain), bricks of graphite (domain) were used in the first (level) nuclear reactor (domain). Other suitable (level) moderators are cadmium (domain) and water (domain). The reactor (domain) is simply (level) a new (level) way to boil water (domain) from the heat (domain) produced (compensation). The boiling (level) water (domain) is used to generate (compensation) electricity (domain). There are several (level) basic (level) designs of nuclear reactor (domain), a common (level) one is the PWR, pressurized water reactor (domain).

A nuclear reactor (domain) does not need (domain) pure (level) U-235 (domain expressed in level) as in a bomb (domain), but it uses uranium (domain) which must be 2-4% (level) U-235 (domain). This is more than (level, comparison) that found in (domain) natural uranium (domain) and so, it must be enriched (level) in much the same way as pure (level) U-235 (domain) is extracted (compensation) for the bomb (domain). Some reactors (domain) use uranium dioxide (domain) as fuel and carbon dioxide gas (domain) at high temperature (level) and pressure (level) as the coolant (domain) to carry away (compensation) the useful heat (domain).

To most (level) of us (domain), the technology (domain) that comes to mind (compensation) first (level) is our cars (domain). I (domain) believe (domain) that quite a bit (level) of credit (domain) is due (domain) to the internal combustion engine (domain). "Experts" (domain) have been predicting (compensation) it's demise (compensation) for many (level) years (domain). In virtually any scenario (domain) of the future (domain), it is conspicuous (level) by it's absence (domain). As far back (level) as my childhood (domain) in the 1960s and early 1970s (domains), I (domain) was (domain) reading (compensation) and hearing predictions (compensation) that the era (domain) of the internal combustion engine (domain) is almost (level) at an end (domain) and will soon (level) join (domain) the steam engine (domain) and black-smith's anvil (domain) in history class (domain). Well, we are at the turn (compensation) of the millenium (domain) and I am still waiting (compensation).

When you turn (compensation) the key (domain) it completes (domain) a circuit (domain) which activates (compensation) a high torque (level) starter motor (domain) under (level) the engine (domain). This is geared (domain) to a large (level) metal flywheel (domain) from which the electric starter motor (domain) is disengaged (domain) after (domain) it begins turning (compensation). The flywheel (domain) is connected (domain) by a crankshaft (domain) to the piston heads (domain) inside (domain) the cylinders (domain) within the engine block (domain). Car engines (domain) vary (level) in number (level) and arrangement (level) as well as size (level) of cylinders (domain).

On the front (level) of the engine (domain), a pulley (domain) at the front end (level) of the crankshaft (domain) operates (compensation) drive belts (domain) as the crankshaft (domain) turns (compensation). One (level) of which (domain) drives (compensation) a fuel pump (domain) bringing fuel (domain) to a carburetor (domain) on top (level) of the engine (domain) as the crankshaft (domain) turns (compensation). New (level) cars (domain) use fuel injectors (domain) instead of (domain) carburetors (domain). In the old (level) carburetors (domain), the right amount (level) of fuel vapor (domain) is mixed with (domain) in-flowing (level) air (domain) after (domain) having passed through an air filter (domain). The fuel (domain) to air (domain) ratio (level) is made richer or leaner (levels) in fuel vapor (domain) as required (comparison, compensation) by a butterfly valve (domain). As the engine (domain) warms up (level), the heat (level expressed in domain) enables the warmer air (domain expressed in level) in the intake manifold (domain), where the fuel-air mixture (domain) is held (domain) before entering (domain) the cylinders (domain), to hold more (level) fuel vapor (domain). This is why cars (domain), especially older (level) models with carburetors (domain) instead of fuel injection (domain), tend to stall (compensation) on cold (level) mornings (domain). The still-cold air (domain expressed in level) in the intake manifold (domain) cannot hold (domain) enough (level) vaporized gasoline (domain) to keep (domain) the engine running. Carburetors (domain) and fuel injectors (domain) usually compensate by producing (compensation) a rich (level) fuel-air mixture (domain expressed in sub-domain) until the engine (domain) warms up (level). The vacuum (domain) produced by piston movement (compensation) draws the air both before and after it is mixed with (domain) fuel vapor (domain) into the engine (domain) just as your lungs (domain) produce vacuum by expanding (compensation) the chest (domain) to draw (compensation) air (domain) in (domain).

The pistons (domain) inside the cylinders (domain) operate in a four-stroke cycle (domain). So if an engine (domain) has four (level) cylinders (domain), one (level) will be in each operational stroke (domain) at any given moment (domain). The four strokes are-intake of the fuel and air mixture, compression of the fuel and air mixture, power upon ignition of the mixture, exhaust of the burned gases (all domains). In the course of these four strokes (domain), a piston (domain) goes through two cycles (domain) with an up (level) and a down (level) movement (compensation) in each cycle (domain). The exhaust (domain) exits (compensation) through an exhaust manifold (domain). The power stroke (domain) is the only one (level) which actually produces (compensation) power (level) but the momentum (level) of the flywheel (domain) keeps the engine (domain) going (compensation) during those moments (domain) when power (level) is not actually being produced (compensation).

Aside from the crankshaft (domain) there is another (domain), smaller (level) shaft (domain) running through the engine (domain) known as the camshaft (domain). This shaft (domain) is driven by (compensation) the crankshaft (domain) and has the function of opening (compensation) valves (domain) to the cylinders (domain) at the right time (domain). Each cylinder (domain) in the engine (domain) has two (level) valves (domain) operated by the crankshaft (domain), one opening (compensation) from the intake manifold (domain) and the other (domain) opening to the exhaust manifold (domain). The intake valve (domain) is opened (compensation) just as the piston (domain) is moving (compensation) away from (level) the valve (domain) in the cylinder (domain), this causes the fuel-air mixture (domain expressed in sub-domain) to be pulled into (compensation) the cylinder (domain) by vacuum (domain), just as air (domain) is pulled (compensation) into your lungs (domain). The exhaust valve (domain) is opened (compensation) just as the piston (domain) is moving (compensation) toward (level) the valve (domain) in a cylinder (domain) and so, it pushes (compensation) the now-burned exhaust gases (domain) out into the exhaust manifold (domain). This coordination (level) requires that the connection (domain) between crankshaft and camshaft (domains) be very precisely set (level). This connection is by a belt or chain (domains) and the setting is known as the timing of the engine (domain).

Now that the engine (domain) is running (domain), we have to be able to make use of the rotary power (level) being generated (compensation). This is where the transmission (domain) comes in. The transmission (domain) can be either manual, using gears, or automatic, using fluid (domains). The transmission (domain) of an automobile (domain) is the link (domain) between the crankshaft (domain) in the engine (domain) and the driveshaft (domain) that turns the wheels (domain). The transmission (domain) makes it possible (domain) for us to control (domain) whether the car (domain) goes forward (level) or in reverse (level) by switching gears (domain) to change the spin direction (level) of the drive shaft (domain). The transmission (domain) also controls (domain) the mechanical ratio (level) of the engine speed (level) to the turning speed (level) of the drive shaft (domain). On standard transmissions (domain), this makes it necessary (domain) for the driver (domain) to shift gears (domain) when the vehicle (domain) speed (level) changes (compensation). An automatic transmission (domain) is so called because this is done automatically (compensation).

So, now we are cruising (compensation). But, not for long (level). The metal parts (domain) of the engine (domain) moving against (compensation) each other (domain) at such speed (level) and force (level) would not take long (level) to seize up and destroy (compensation) the engine (domain). Especially since metal (domain) expands (compensation) when heated (level). The heat (domain) produced (compensation) in the cylinders (domain) combined with (domain) that of friction (level) would be devastating (compensation). That is where oil (domain) comes in, the lifeblood of the engine (domain). Oil lubrication (domain) operates on the principle that friction (level) is much less (level) in liquids than in solids (level). Circulating oil (domain), passed through a filter (domain) to cleanse (compensation) any particles (domain) drastically reduces (level) friction (level) and absorbs (compensation expressed in domain) much of the excess heat (level).

But still, an engine (domain) produces enough heat (level) to destroy (compensation) itself (domain) in short order (level). So, a cooling system (domain) which transfers (compensation) heat to a radiator (domain) to be dissipated (compensation) by in-rushing (level) air (domain) is included (domain). Some cars (domain) like Volkswagen Beetles (domain) and Chevrolet Corvairs (domain) have used air alone (domain) as a coolant without any liquid anti-freeze (domain). This left the car (domain) with one less (level) system (domain) to go wrong but with more noise (level) without the muffling effect (compensation) of antifreeze (domain).

An engine (domain) with one hundred percent (level) efficiency (level) would produce (compensation) no waste heat (level). However, such a device (domain) does not exist and some heat (level) is desirable (domain) to get the oil (domain) flowing in cold weather (domain expressed in level). This is why starting a car (domain) in cold weather (domain) puts the most (level) strain (compensation) on an engine (domain), the oil (domain) has not yet begun (domain) to flow to lubricate the engine parts (domain).

Once the car (domain) is moving (domain), it is nice (domain) to be able to control (domain) which direction (level) the car (domain) is going (compensation). A steering wheel (domain) is placed to control (domain) the bearing (level) of the front wheels (domain). The steering system (domain) works either by a system of gears (domain) or a rack and pinion (domain).

Sooner or later (levels), we will need to stop (compensation) our car (domain). This is done by pressing (compensation) a pedal (domain) alongside (level) the accelerator (domain) which causes either a caliper device (domain) to grip (compensation) a disc (domain) inside (domain) the wheels (domain) in disc brakes (domain) or, two crescent-shaped (level) shoes (domain) to expand (domain) outward (level) gripping (compensation) the inside (domain) of a drum (domain) in drum brakes (domain). Unlike inside the engine (domain) where friction (level) is an enemy (domain) to be fought with oil lubrication, the brakes (domain) produce (compensation) great friction (level) to stop (compensation) the car (domain). Disc brakes (domain) are more efficient (level) but also more expensive (level) so, they are often (level) used on the front wheels (domain) where the strain of braking (compensation) is greatest (level), while drum brakes (domain) are used on the rear wheels (domain).

At high speeds (level), the car (domain) is significantly (level) slowed (compensation) by air (domain) resistance (compensation). Automotive designers (domain) try to create (compensation) an aerodynamic shape (level) while pleasing the eyes (domain) and not sacrificing too much (level) interior space (domain). If roads (domain) were always dry (level), smooth (level) tires (domain) would be most efficient (level) as this would provide the maximum (level) surface contact (domain) with the asphalt (domain). But since it may rain (compensation) while we are driving (domain), tires have various (level) tread patterns (domain) to channel away (compensation) water (domain).

PRIMES IN HUMAN AFFAIRS

As we have seen, exactly the same primes are manifested in science, technology and, human affairs. In this section, we will get used to seeing a slice of human history in terms of primes.

The city of Jerusalem (domain) has been playing a monumental (level) role (domain) in human history (domain) for thousands (level) of years (domain). In the year 70 A.D. (domain), the Romans (domain) expelled (compensation) the Jews (domain) from their homeland of Palestine (domain) as punishment (compensation) for an uprising (compensation expressed in domain) against Roman rule (domain), Jerusalem (domain) was already (level) a very old (level) city (domain) at this time (level).

In 632 (domain), Jerusalem (domain) was conquered (compensation expressed in domain) by Arabs (domain) and was renamed Al-Quds (domain). The Arabs (domain) already considered (domain) Jerusalem (domain) as a holy place (domain). On the mount (domain) where the Temple of the Jews (domain) once was (level), The Arabs (domain) completed (compensation expressed in level) the Dome of the Rock (domain) in 691 (domain). A large (level) mosque (domain) was built (compensation) next to (level) the dome (domain) a few years (level) later (domain).

The crusaders (domain) from Europe (domain) later (domain) got control (domain) of Jerusalem (domain) and turned (compensation) the Dome of the Rock (domain) into a church (domain). Most (level) crusaders (domain) from the first (level) crusade (domain) went (compensation) home (domain) when Jerusalem (domain) was captured (compensation expressed in domain). Those (domain) who stayed (domain) set up (compensation) a kingdom (domain) in the holy land (domain) and controlled (domain) Jerusalem (domain) for about a century (domain) but war (domain) with the Moslems (domain) never stopped (domain). The second (level) crusade (domain) to defeat (compensation) the Moslems (domain) was ineffectual (level) because the armies (domain) from different (level) countries (domain) did not coordinate (level) their strategies (domain). There were (domain) several (level) separate (domain) crusades (domain) over about two (level) centuries (domain). In 1187 (domain) Saladin (domain), a Kurdish (domain) leader (level), recaptured (compensation expressed in domain) Jerusalem (domain) for the Moslems (domain). Much (level) work (compensation) was done on Jerusalem (domain) by later Moslem (domain) rulers (level). The city (domain) was greatly (level) expanded (domain) when controlled (domain) by the Ottoman Turks (domain).

The Ottoman Empire (domain) was on (domain) the losing (domain) Central Powers (domain) side during World War One (domain) and Jerusalem (domain) fell under British control (domain) from 1917-1948 (domain). The state of Israel (domain) was reborn (domain) at this time (level) and old Jerusalem (domain) with the temple mount (domain) was recaptured (compensation expressed in domain) by the Israelis (domain) in the 1967 six-day war (domain). This was the first time (level) in 1,897 years (level) that the Jews (domain) had control (domain) of their most sacred (level) place (domain).

The primary (level) effect (compensation) of the crusades (domain) was to turn (compensation) European (domain) attention (domain) to the outside (level) world (domain). Ships (domain) brought back (compensation, domain) silks, carpets and, cotton (all domain) that some Europeans (domain) had never (domain) seen before (domain). This was probably (level) the first time (level) Europeans (domain) had tasted (domain) lemons, oranges, cinnamon, pepper and, rice (all domain). This no doubt (level) did much (level) to pave the way (domain) for the great (level) European (domain) explorers (domain) a few (level) centuries (domain) later (domain).

The Ottoman Turks (domain) came into prominence (level) at about the time (domain) the Crusades (domain) were ending (domain) by defeating (compensation) Byzantium (domain). Natives (domain) of the Black Sea coast of Anatolia (domain), The Ottoman Turks (domain) would control (domain) the Moslem world (domain) for six (level) centuries (domain). Their empire (domain) extended (domain) into Europe (domain) as far as (level) the Danube River (domain). The most famous (level) of the Ottoman Turks (domain) was Suleiman the Magnificent (domain), who died (compensation expressed in domain) in 1566 (domain).

After (domain) the reformation (compensation expressed in domain) in Europe (domain), the newly (level) protestant nations (domain) began a march (domain) of progress (compensation) including (domain) invention, scientific discovery, exploring new lands and, the industrial revolution (all domains). This put (compensation) the Ottoman Turks (domain) on the defensive (domain). A large (level) Ottoman army (domain) was devastated (compensation) in an attempt (domain) to capture (domain) Vienna (domain). European nations (domain) began (domain) taking (domain) Ottoman-controlled territory (domain) as their own colonies (domain). France (domain) took control of Algeria and Tunisia, Britain took Cyprus and Egypt (all domains). The Ottoman Empire (domain) was finished (compensation expressed in domain) by being on the wrong side (domain) in World War One (domain).

Early (level) in the twentieth century (domain), notice was served (compensation) that east Asia (domain) was going to be (domain) a major (level) player (domain) in the new (level) century (domain), not as European colonies (domain) but in it's own right (domain). In 1905 (domain) Russia (domain), under the leadership (domain) of the czars (domain), suffered an unexpected (domain) and devastating (compensation) naval (domain) defeat (compensation). The fleet (domain) of Japan (domain) had the Russian pacific fleet (domain) beseiged (compensation expressed in domain) at Port Arthur (domain), on the Asian mainland (domain).

Russia (domain) had another (domain) naval fleet (domain) in the Black Sea (domain). But, the fleet (domain) was needed there (domain) in case of conflict (compensation) with Ottoman Turkey (domain). And hostile (level) Britain (domain) controlled (domain) Gibraltar (domain) and the Suez canal (domain), one of which the Russian fleet (domain) would need (domain) to pass through to reach east Asia (domain). The Russians (domain) decided (comparison) to send (compensation) it's far-distant (level) Baltic sea fleet (domain) on a seven-month (level) journey (compensation expressed in domain) around the continent of Africa (domain) to break (compensation) the Japanese seige (domain) of Port Arthur (domain).

However, the Russian sailors (domain) were exhausted (level) by the long (level) journey (compensation expressed in domain) and were surprised (domain) by Japanese ships (domain) waiting for them between Japan and Korea (domains). The Russians (domain) were devastated (compensation) in a short (level) battle (domain). This defeat (domain) deeply (level) affected (compensation) morale (level) in St. Petersburg (domain) and doubtless (level) contributed (domain) to the brewing uprising (compensation expressed in domain) against the czars (domain). In Japan (domain), it certainly bred (compensation) the confidence (level) to launch it's attempt (compensation) to bring all of Asia (domain) under it's control (domain) in the 1930s (domain).

Not many (level) countries (domain) have been around (domain) as long (level) as Egypt (domain). It's fortunes have waxed and waned (levels) as endless centuries (domains) flew by. But the Egypt (domain) of the Pharaohs, the pyramids and, the sphinx (all domains) is still here (domain) for yet another millenium (domain).

In the nineteenth century (domain), attention focused (domain) on Egypt (domain) for two (level) reasons (domains). Archeology (domain) was on it's way (level) to becoming a modern (level) science (domain) and nowhere (domain) was more significant (level) than Egypt (domain). If there were a sea route (domain) between the Mediterranean Sea (domain) and the Red Sea (domain), it would not be necessary (domain) to sail around Africa (domain) to reach Asia (domain) and the world (domain) would seem like a much smaller (level) place (domain). The desire (domain) for a maritime shortcut (domain) led to the construction (compensation) of the Suez Canal (domain), which was completed (domain) in 1869 (domain). Britain (domain) remained in control (domain) of Egypt (domain) until 1922 (domain). During this time (domain), the original Aswan Dam (domain) on the Nile River (domain) was completed (domain) in 1902 (domain).

Egypt (domain) was ruled (domain) by King Farouk (domain) until 1952 (domain) when he was overthrown (compensation) by a nationalist (level) army (domain) officer (level), Gamal Abdel Nasser (domain). Dissatisfaction (level) with the poverty (level) of peasants (domain) and the 1948 (domain) defeat (compensation) by Israel (domain) were the catalysts (domain) for Nasser's (domain) rise (compensation) to power (level).

Nasser (domain) was a socialist (domain) and undertook (compensation) extensive (level) land (domain) reforms (compensation). He aroused (compensation) dislike (domain) in the west (domain) for shunning (compensation) democracy (domain). He nationalized (domain) the Suez Canal (domain) and got an unexpected (domain) military (domain) response (compensation) from Britain (domain), France (domain) and, Israel (domain). But by this time (level), the era (domain) of empires (domain) was at an end (domain) and the United Nations (domain) pressured (compensation) the three (level) to withdraw (compensation). Nasser's people (domain) looked upon him (comparison) as a hero (level).

In 1958 (domain), Egypt (domain) and Syria (domain) formed a union called "the united Arab Republic" or "UAR' (domain). But Syria (domain) began to feel that it was an unequal (level) partnership (domain) and wealthy (level) Syrians (domain) disliked (domain) Nasser's socialist thinking (domain) and Syria (domain) withdrew (compensation) from the UAR (domain). This is why some maps (domain) refer to Egypt (domain) alone as "the UAR' (domain). Nasser (domain) received (domain) loans (domain) from the Soviet Union (domain) to build (compensation) the Aswan High Dam (domain) on the Nile River (domain) for irrigation (domain) and hydroelectricity (domain).

Nasser (domain) died (compensation expressed in domain) in 1970 (domain) after seeing Egypt (domain) defeated (compensation) in the 1967 six-day war (domain). Almost all (level) of the Egyptian air force (domain) had been destroyed (compensation) on the ground (domain) before (domain) it could get into (compensation) the sky (domain) to fight (compensation). Anwar Sadat (domain) took (compensation) power (domain) upon Nasser's death (domain).

Sadat's Egypt (domain) was also defeated (compensation) by Israel (domain) in yet another (domain) middle east war (domain), the Yom Kippur war (domain) in 1973 (domain). This time (domain) however, everyone (domain) knew (domain) that Israel (domain) had a very close win (level). Both (domain) the Israelis (domain) and Arabs (domain) became more willing (level) to seek peace (domain). Sadat (domain) is best known (level) in the west (domain) for signing (compensation) the peace treaty (domain) with Israel's (domain) Menachem Begin (domain) at Camp David (domain). The negotiating (domain) was mediated (compensation) by Jimmy Carter (domain). Sadat (domain) pleased (compensation) millions (level) of people (domain), and infuriated (compensation) millions (level) of others (domain), by visiting (compensation) Jerusalem (domain).

In 1981 (domain), Sadat (domain) was assassinated (compensation) by Egyptian (domain) hard-liners (domain) wanting nothing to do with peace (domain) with Israel (domain). Hosni Mubarak (domain), Sadat's vice-president (domain) took (compensation) power (domain). Mubarak (domain) has continued (compensation) Sadat's (domain) policy of peace (domain) with Israel (domain) despite the displeasure (domain) of millions (level) of Moslems (domain). Mubarak (domain) was once (domain) a military pilot (domain) and was in charge (domain) of the Egyptian air force (domain) until Sadat (domain) appointed (compensation) him (domain) as vice-president (domain) in 1975 (domain).

In the new world (domain), there has also (domain) been exciting (level) stories (domain) of nations (domain) and leaders (domain). Take Argentina (domain) for example. Around 1800 (level), people (domain) in what is now Argentina (domain) had considerable (level) autonomy (domain) under loose (level) Spanish control (domain). Spain's center of control (domain) in

South America (domain) was distant Lima (domain), in what is now Peru (domain). Regions like Argentina (domain) were on the fringe (level) of Spain's vast empire (domain). Argentina's name (domain) means "the land of silver" (domain) but, it did not have (domain) the gold (domain) that the Spaniards (domain) found (compensation) in Peru (domain).

In Europe (domain), a new (level) conqueror (domain) emerged (compensation). Napoleon (domain) conquered Spain (domain), cutting the empire (domain) off (compensation) from it's colonial master (domain). This is what gave (compensation) momentum (level) to independence movements (domain) in South America (domain) as well as in places (domain) such as far away (level) Mexico (domain). During Napoleon's occupation (domain) of Spain (domain), residents (domain) of Argentina (domain) declared independence (compensation). After the demise (compensation) of Napoleon's empire (domain) in Europe (domain), Spain (domain) was defeated (compensation) in an attempt (compensation) to reassert control (domain) in Argentina (domain).

One of the most famous (level) Argentines (domain) is Jose de San Martin (domain), who continued the struggle (compensation) for South American independence (domain) after Argentina (domain) had secured independence (domain). San Martin (domain) led (compensation) an army (domain) westward (level) to Chile (domain) and then north (level) to Ecuador (domain). San Martin (domain) met (compensation) South America's other great liberator, the Venezuelan Simon Bolivar (domain), but the two (level) had a disagreement. San Martin (domain) reportedly forgot (domain) about South America (domain) and moved (compensation) to France (domain) for the remainder (level) of his life (domain).

Argentina (domain) is best (level) known (domain) not for precious metals (domain) but for beef (domain). A new method of freezing (compensation) meat (domain) was developed (compensation) in the late 1800s (domain) allowing Argentine beef (domain) to be shipped (compensation) to Europe (domain), where it usually sold out (compensation) quickly (level). Land (domain) was plentiful (level) and visions of Argentine ranching (domain) attracted (compensation) millions (level) of European Immigrants (domain). Unlike (domain) other South American nations (domain), Argentina (domain) has never paid much (level) attention (domain) to it's Indian heritage (domain) and remains to this day (level) as the most European (level) country (domain) in South America (domain).

Chapter Eight

▼

Applications and Ramifications of Primes

A NEW WAY OF LOOKING AT THE UNIVERSE

I expect that the first change resulting from the Theory of Primes will be in our way of thinking about the universe. Does it not make you feel a little bit differently to know that there is only four things that can exist or can happen and that everything consists of manifestations and combinations of these four things? The entire universe and everything that exists or can exist seems suddenly to be very simple. Domains, levels, comparisons and, compensations is all there is.

The advantage of having a new way of looking at everything is simply that it will lead us to notice things that we did not notice previously. Every inventor and writer knows that this jolt out of grooved-in thinking patterns is very valuable. I believe that this will lead to new discoveries by promoting people to notice things that were not noticed before. Some of these new discoveries may prove very useful, whether in the realms of science and mathematics or philosophy.

EDUCATION

Every idea and concept can be broken down into four components. The value of this to the education community is incalculable. If students can master the postulates of primary theory, which governs every topic and subject, learning should come much easier. Any subject matter can be difficult to grasp if the student does not have a command of the basic principles and becomes easier when the principles are understood. In all of reality, the most basic principles are the primes.

CATEGORIZING

Now that we can break anything down into the four primes, it is possible to express anything that exists or happens quickly and easily. The more we know of that items have in common and the more that the items can be broken down, the easier it is to categorize the items. The primes are an infinite and eternal measuring tool for anything that exists.

It is very beneficial that primes are being discovered during the computer revolution. Using the primes and a computer program, it is possible to categorize everything that can possibly be by categorizing domains with appropriate levels, then the appropriate comparisons and possible compensations that may result. Just put the data for primes in a mainframe computer and see what it can come up with.

The primes are an entirely new method of categorization for everything and a program structured by primes can discover possibilities that would not be thought of otherwise. Just as computer hackers use "brute force" programs to try every possible password, programs structured around the primes can come up with every possible combination of inputted data. Based on the primal fact that domains, levels, comparisons and, compensations is all there is.

SCIENCE AND MATHEMATICS

The Theory of Primes upsets the status quo in a very important way. Science and mathematics are closely intertwined. Science has usually made the discoveries and mathematics quantified, described and, provided a model for developments in science. In other words, science usually led the way and mathematics followed. Science built the road and mathematics put the signs along the road. The ancients did not develop geometry and trigonometry until there was centuries of experience in building everything from simple houses to temples. The "real world" came first and mathematics followed.

The Theory of Primes has changed all of that. For the first time since antiquity, mathematics is now leading science. Primes are a newly discovered branch of mathematics that is more primal than the counting numbers that have been in use for thousands of years. Primes define and set the forms for what can exist, including everything that scientists can contemplate or discover. This means that science is now leading mathematics by defining what can exist rather than merely describing and quantifying it.

I went back to college in my thirties and used to sit in class with a vague feeling that something was backwards or that something was missing. I did not quite grasp it at the time but it was both. Something vital was missing. Mathematics had it's keystone missing. Primes are the most fundamental branch of mathematics and with this branch, mathematics moves in front of science by defining everything that can exist rather than just quantifying and describing the findings of science.

MATHEMATICS AND LANGUAGE

At this point, the primes will be used for expression rather than calculation like other branches of mathematics are. However, I am certain that it will not take long for many new uses to be found for the primes.

The primes can give us a mathematical accuracy that is never possible with question words like who, what, when, where, why, which, how, how much and, how many.

The words; who, what and, which are domain related. When, where, how much and, how many involve levels. How and why concern comparison and compensation. These words do illustrate that there are certain fundamental patterns that seem to govern everything. But it is not possible to understand or to get any use from the four basic primes using awkward, inexact and inefficient wording such as this. The primes and the precision use of primes belongs in the domain of mathematics and not semantics or language. Words in the dictionary could be categorized as to which prime the words fall under.

The primes are very much a boundary region. Not only is the primes the most primary branch of mathematics, it is also where language meets mathematics. The symbols of mathematics actually are a language but it was always believed that it was a different realm than the English language. The region of primes, which are a concise and precise method of expressing the same concepts as the question words, is the previously uncovered boundary region where mathematics meets everyday language.

The meeting of language and mathematics at the primes is best illustrated in manifestations of the level prime. Numbers are in the domain of mathematics while verbal descriptions are in the domain of language. Expressed in primes, both are levels.

Whenever the explorers of hundreds of years ago discovered a previously unknown region that bordered two known but separate regions, it brought geographers a step closer to mapping the entire world. The primes are the most fundamental branch of mathematics and border the realm of language. This means that the discovery of primes is a major step to the "mapping" of reality.

What is different than before the Theory of Primes is that the patterns we see in the world all around us as well as the question words have been broken down into the four basic components. These components, the primes, are the basis for all that exists or ever can exist. Have you ever thought about the question words and wondered how many basic "patterns" there are which govern everything? Have you ever wondered while doing math homework if mathematics, which seems to be able to quantify everything that exists, had a branch that could define not just what could happen but what could exist? The answer to both questions is the four primes.

THE THEORY OF EVERYTHING

Since I am claiming that the Theory of Primes is the best answer to the search that is currently under way in the scientific community for a theory of everything, it is appropriate to take a closer look at the objectives and background of this search for a theory of everything.

As it stands now, the electromagnetic force, the strong nuclear force that binds atomic nuclei and, the weak nuclear force that causes radioactivity may be open to being brought together. The electromagnetic and weak forces especially, since these two have been shown to act on quarks and leptons equally and, quarks and leptons are the fundamental components of matter in what is known as the quark model. Some physicists call a combination of the two the "electroweak force". The strong nuclear force has proven more difficult to include although at extremely high energies and temperatures in particle accelerators the three forces begin to show a lot in common.

Many people, including me, do not consider gravity as a genuine force but rather an innate property of space itself. So, there is no reason to be concerned about including gravity in the unification of forces. It is believed that there are tiny particles that act as "force carriers" for the forces. Particles called "bosons" are thought to transmit the electromagnetic, weak and, strong forces. While some think that "gravitrons" transmit gravity. If this is the case, it would mean that gravity is a force like the other three and not an innate property of space itself.

Quantum mechanics is a mathematically expressed concept developed by physicists between the world wars. It explains the operation of the electromagnetic force, the strong nuclear force and, the weak nuclear force. Einstein spent quite a bit of his life trying to develop a theory that would explain both gravity, which Einstein postulated is caused by warping of space, forces and, sub-atomic particles. This was to be called the "unified field theory". Reports are that Einstein was not too thrilled by the ideas of quantum mechanics, answering that "God does not play dice" in response to the statistical predictions that quantum mechanics attempts to apply to sub-atomic particle behavior.

One thing that many theoretical physicists do agree on is that the secret to a theory of everything most likely lies at the events comprising the big bang, the beginning of the physical universe. All forces may have briefly existed as one in a condition of "symmetry". It is believed that as the universe exploded outward, the forces broke off from each other and so became separate as it is today.

The Theory of Primes drives home like nothing else what everything in this universe, as well as what anything that is outside the universe, has in common. I can think of no better unifying revolution in thought than the common structure of everything in existence, no matter how completely different the things seem to us. After finding the primes, it now seems to me as if everything that can exist is in invisible containers akin to mason jars that we had not noticed previously. No matter where you go or what you do, the same jars will be there. The only difference is what is in the jars. The jars are of course, the primes. what is in the jars is all that exists.

Science has broken down the universe into all manner of laws to guide and set limits the operation of just about everything that exists. The primes are the laws for what can exist to begin with.

There has been tremendous advances in science in the twentieth century. But no matter what is discovered or invented, exactly the same four primes are manifested. As so many of the things that were believed at the beginning of the century were overthrown. As our little earth has been shown to be more and more insignificant in the universe, the primes have always remained the same even though not discovered until now.

Even though scientific research gets ever more specific rather than general, there was a feeling that there is something that serves as a great unifier. This something was the object of the search for the theory of everything. It was believed that the basic forces could be linked to some common source or particle and possibly that Einstein's theory of relativity and quantum mechanics could be linked together.

Whatever results from the physicists' search for the theory of everything, the search is only taking place in the physical universe, the given, and will only find whatever answers there are in the physical universe. The search for a theory of everything as it stands now does not include anything to do with human beings such as economics, politics or, chocolate and caffeine addiction. My Theory of Primes encompasses everything that exists or can exist.

There is a simple formula, a four-part sequence for everything there is.

REALITY=DOMAIN>LEVEL>COMPARISON>COMPENSATION

I do not know the common denominator of all the basic forces, and I will leave the search for it up to the physicists searching for the theory of everything. All I know is that whatever it is, it is part of the given and is defined by the primes like everything else.

The search for the theory of everything is taking place in the realm of science, even though mathematics will certainly be used in it's description whenever the answer is found. However, since my discovery of primes has altered the status quo and put mathematics ahead of science, does it not make sense that the search for the theory of everything should now concentrate on the realm of mathematics rather than science? I believe that I have found the theory of everything in the mathematical realm and I have named it the Theory of Primes.

THE TRINITY OF PRIMES

There is an enormous religious significance to the discovery of primes. As I mentioned earlier, I have chosen to present four primes. The primes are easier to explain if comparison and compensation are considered separately. Also, when making practical use of the primes, such as writing a computer program, comparison would have to be used separately. However, comparison and compensation go together. Comparison makes no sense unless it is done for the purpose of bringing about an appropriate compensation, if necessary, which cannot happen without a comparison preceding it. The result of this is that we could say there is three, rather than four, primes.

We can see the logic of having three primes instead of four when we consider that comparison is unlike domain and compensation in that it cannot be expressed in levels. Domain and compensation can be expressed in levels and levels can be expressed in domain. The interexpressibility of the primes aside from comparison lead one to believe that comparison is not a real prime but merely the door to compensation.

It cannot be disputed that comparison is necessarily included in the primes and that it is easier to explain primes when considering comparison as separate from compensation. When primes are used to categorize, comparison must be a separate step in the process to be effective. Yet, we can see by the illustration of interexpressibility that there is reason to not classify comparison as a separate prime. What could be done is to compromise and refer to comparison as a "vital entity" in the primes rather than as a full prime. We could also say that comparison is a sub-prime of compensation.

POSTULATE 17: COMPARISON MAY BE CONSIDERED AS A FULL PRIME OR AS A SUB-PRIME OF COMPENSATION OR AS A VITAL ENTITY IN THE OPERATION OF THE PRIMES

If comparison is not considered as a full prime, then we have three or, a trinity of primes. The God of Christianity is also a trinity. The Father, the Son and, the Holy Spirit. Christians have noticed that much of the world and the universe can be broken down into a trinity. A person who makes something, whether painting or building, leaves something of himself in his work.

So, we could expect that since God created everything, something of God is to be found everywhere.

Sure enough, God is a trinity and this pattern is to be seen in many aspects of the universe. The universe consists of a trinity-space-time, matter and energy. Space has three dimensions. There are three forces-the electromagnetic force, the strong nuclear force and the weak nuclear force. (According to Einstein's theories, gravity is not a genuine force but an innate part of space and the so-called fifth force is speculation at this point). There are three basic components of atoms-protons, neutrons and, electrons. There are three basic types of galaxy-spiral, elliptical and, irregular. Radioactivity comes in three forms-alpha, beta and, gamma.

The earth consists of three basic parts-land, water and, air. There are three basic types of cloud-cirrus, stratus and, cumulus. There are three basic layers to the earth's atmosphere-the troposphere, the stratosphere and, the ionosphere. There are three basic layers to the earth-the core, the mantle and, the crust.

A human being is a trinity-the mind, the body and, the spirit. There are three basic races of humans-white, black and, oriental. Human history can be best divided into three parts-ancient times, the middle ages and, modern times. The fundamental family unit consists of husband, wife and, child.

Now, there are three primes which define everything that exists or can exist. This is of course, if we count comparison and compensation as one prime. I have introduced four primes, with comparison and compensation as two primes only for ease of understanding. This appears as powerful evidence for the veracity of Christianity as well as for the Theory of Primes.

Appendix A

---------------▼---------------

LIST OF POSTULATES

POSTULATE 1: PRIMES DEFINE WHAT CAN EXIST AND WHAT CAN HAPPEN. EVERYTHING IS MANIFESTATIONS OF AND COMBINATIONS OF THE PRIMES.

POSTULATE 2: THE GIVEN MANIFESTS PRIMES

POSTULATE 3: PRIMES ARE THE MOST FUNDAMENTAL COMPONENT OF PHYSICAL REALITY

POSTULATE 4: PRIMES ARE MATHEMATICAL ENTITIES WHICH ARE MADE REAL BY MANIFESTATION.

POSTULATE 5: THE FIRST PRIME IS DOMAIN AND IS MANIFESTED WHENEVER ANYTHING EXISTS.

POSTULATE 6: THE SECOND PRIME IS LEVEL AND IS ALWAYS MANIFESTED BY DOMAIN. AN EXPRESSION CONCERNING DOMAIN IS LEVEL.

POSTULATE 7: PRIMES CANNOT EXIST IN LEVELS. NO MANIFESTATION OF ANY PRIME IS ANY MORE OR LESS A MANIFESTATION OF THAT PRIME THAN ANY OTHER MANIFESTATION OF THAT PRIME. ONLY MANIFESTATIONS OF A PRIME CAN MANIFEST LEVELS.

POSTULATE 8: DOMAIN IS DEFINED BY BOUNDARY WHILE LEVEL IS DEFINED BY BARRIER. A BARRIER REQUIRES SOME MANIFESTATION OF LEVEL TO CROSS WHILE BOUNDARY DOES NOT.

POSTULATE 9: PRIMES MUST BE MANIFESTED IN SEQUENCE. A CONDITION IS A DOMAIN. A CONDITION OF A CONDITION IS A LEVEL. ANYTHING THAT A NUMBER OR DESCRIPTION CAN BE ATTACHED TO IS A DOMAIN.

POSTULATE 10: THE ONLY PRIME THAT CAN MANIFEST SUB-PRIMES IS DOMAIN. THERE CAN BE NO SUB-LEVELS BECAUSE DOMAIN IS THE FIRST PRIME AND ANY MANIFESTATION OF SUB-LEVELS WOULD NECESSARILY BE DUE TO SUB-DOMAINS.

POSTULATE 11: THE THIRD PRIME IS COMPARISON, WHICH COMPARES LEVELS.

POSTULATE 12: THE FOURTH PRIME IS COMPENSATION, WHICH ADJUSTS LEVELS TO SEEK STABILITY ON INSTRUCTION FROM COMPARISON.

POSTULATE 13: ALL CHANGES OF ANY KIND ARE MANIFESTATION OF COMPENSATION AND ARE EXPRESSIBLE IN LEVELS.

POSTULATE 14: MANIFESTATION OF COMPENSATION IS A DOMAIN IN ITSELF. THIS CONNECTION FROM THE FOURTH PRIME BACK TO THE FIRST PRIME IS CALLED THE PRIME CYCLE.

POSTULATE 15: ALL COMPENSATIONS OVERCOME A BARRIER.

POSTULATE 16: THE ONLY DIFFERENCE BETWEEN ALL DOMAINS AND ALL COMPENSATIONS IS LEVELS.

POSTULATE 17: COMPARISON MAY BE CONSIDERED AS A FULL PRIME OR AS A SUB-PRIME OF COMPENSATION OR AS A VITAL ENTITY IN THE OPERATION OF THE PRIMES

Appendix B

---▼---

GLOSSARY OF TERMS

ACTIVITY FACTOR-The proportion of comparisons resulting in a compensation. The more compensations, the higher the activity factor. This term could also express the degree of compensations taking place.

ADVENT OF THE GIVEN-In our universe, the big bang brought the given, the physical universe, into being. This term refers to the big bang, which is also known as the primary compensation.

COMBINATION-Primes combining on the same level or different levels. Everything in existence consists of combinations of and manifestations of the primes.

COMPARISON-The third prime. Continuous comparison of levels is always going on to see if any compensation is called for. Comparison and the fourth prime, compensation, may possibly be considered as one prime because the two work together.

COMPENSATION-The fourth prime. If comparison detects a change in levels or, if a state of maximum possible stability is not existent, a compensation may be called for.

COMPENSATION CYCLE-Compensation is the fourth and last prime. Domain is the first prime. But a compensation brings a domain into manifestation, reverting back to the first prime.

CONSCIOUS LEVELS-In our universe, any levels of reality above the base level of inanimate matter. The conscious levels only come into manifestation for a being possessing conscious-ness, such as some people.

DEFINED DOMAIN-A domain defined in reality by a con-scious being. This domain would be meaningless to inanimate matter. An example of defined domain is a meter.

DOMAIN-The first prime. Domain is manifested whenever anything exists. A condition or an object is the most familiar domains to us. Domain is defined by a boundary.

DOMINANCE OF EXPRESSION-There are many instances when something can be expressed in either domain or level. Sometimes the manifestation is more a level then domain and vice versa. Warm is a level, water is a domain, warm water would probably be considered more in domain than level.

DOUBLE DOMAIN-Seen when dealing with conscious beings. A compensation results in two domains rather than one. When someone gains interest in soccer, the domain of soccer fans is affected and the individual's mind adds another sub-domain.

EVENT-In our universe, an event is something that happens. An event is a compensation and forms a domain. This completes the cycle of primes.

FORCES-In our universe, the strong nuclear force binding atomic nuclei, the weak nuclear force resulting in radioactivity and, the electromagnetic force. Some people consider gravity as a force rather than an innate part of space-time. Others believe that there is a mysterious fifth force that weakly counteracts gravity. The theory of everything is a not-yet-found common source of all the forces. The forces have nothing to do with the primes.

GIVEN-That which manifests the primes. In our universe, the given is space-time, forces matter and, energy. The primes are the same regardless of the given.

INANIMATE LEVEL-The lowest, most common level of reality. Primes or other mathematical entities are manifested by reality, which has different levels. The levels of reality brought about by consciousness are all that is above the inanimate level. We could therefore assume that reality is shaped like a pyramid.

INTEREXPRESSIBILITY-Domain and level can be expressed in each other. Compensation can be expressed as domain or level. Comparison is not interexpressible.

LEVEL-The second prime after domain. A condition is a domain, a condition of a condition is a level. Manifestation of levels inevitably follows manifestation of domain.

LEVELS TRADEOFF-An exchange of levels often seen in our universe. Resulting from the fact that there is only a fixed amount of given. Such a change of levels is a compensation. An example would be money vs possessions, when one goes up, the other goes down.

MANIFESTATION-That which makes a prime or other mathematical entity real. The number six is not real until it is manifested by something such as a six-pack of Coca-Cola. The domain prime does not become real until it is manifested by something such as a solar system.

MATHEMATICS-A collection of entities defining, describing and manifested by reality. Primes, numbers and, geometric shapes are mathematical entities.

NEGATIVE COMPENSATION-Every action is a compensation. A negative compensation is an issue of the given. It prevents, rather than causes, any change in levels.

NEGATIVE EVENT-Something which does not happen. A continuation of the status quo with no compensation taking place. Comparison saw no change in levels requiring a compensation. When a compensation does take place, it is referred to as a positive event.

PATTERNS-Complex primes such as pyramid, pendulum, series, alphabet, seed, bridge and, path. All patterns consist of and can be broken down into primes. Patterns are dependent on the given for their very existence while primes are dependent on the given only for manifestation.

POSITIVE EVENT-A compensation which is manifested in the given as something that does happen. There is a change in levels.

PRIMARY COMPENSATION-In our universe, the big bang, which began the compensation cycle as we know it.

PRIMARY THEORY-The Theory of Primes is a part of the new field of primary theory. The study of that which is most primary in physical reality, the basic patterns defining what can exist and can happen.

PRIME-The four mathematical entities which define everything which can exist or can happen.

PRIME CYCLE-The last prime is compensation. But when compensation is manifested, it also manifests the first prime, domain.

REALITY-That which manifests primes, also known as the given.

RELEVANCE-Due to the nature of the given, comparisons only compare and compensations only affect the appropriate levels. For example, when you ring a doorbell, it causes a bell or chime in the house in which the doorbell is ringing. It does not cause the sound of a bell to emanate from a rock on Mars because it would not be relevant.

STABILITY-Due to the nature of the given, comparison seeks the lowest energy level. This is why objects fall.

STATUS QUO-Things as they are. Comparison deciding on no compensation.

SUB-DOMAIN-A domain which is part of a larger domain on a particular level. Domain is the only prime which manifests sub-primes. There are no sub-levels since any sub-levels would be brought about by sub-domains.

THEORY OF EVERYTHING-The search for a common source for the basic forces of the universe. Also encompasses the search for a link between quantum theory and relativity. A theory to explain everything.

UNIT-A domain defined by a human or other conscious entity which would have no real meaning in the inanimate level of reality. This is a flexible term but, I define unit as meter, inch, yard, etc.

VIEWPOINT DOMAIN-A domain formed by the attention of a conscious entity. A viewpoint domain would not ordinarily be manifested without the conscious entity.

VITAL ENTITY-The comparison prime may be considered as part of the compensation prime rather than as a prime in itself. In this case, we could say that comparison is a vital entity rather than a full prime.

About the Author

▼

Mark Meek is from the village of Lydbrook in the Forest of Dean in Gloucestershire, England and now lives in America. He has a life-long fascination with science, particularly physics and astronomy, and was convinced that there exists something contributing to orderliness in the universe yet to be discovered.

www.ingramcontent.com/pod-product-compliance
Lightning Source LLC
Chambersburg PA
CBHW030757180526
45163CB00003B/1058